RealTime Physics
Active Learning Laboratories

Module 2

Heat and Thermodynamics

The production of motion in a steam engine is always accompanied by a circumstance which we should particularly notice. This circumstance is the passage of caloric from one body where the temperature is . . . elevated to another where it is lower.

—S.N.L. Carnot (1824)

David R. Sokoloff
Department of Physics
University of Oregon

Ronald K. Thornton
Center for Science and Math Teaching
Departments of Physics and Education
Tufts University

Priscilla W. Laws
Department of Physics
Dickinson College

John Wiley & Sons, Inc.
New York • Chichester • Weinheim • Brisbane • Singapore • Toronto

ACQUISITION EDITOR   Stuart Johnson
MARKETING MANAGER   Catherine Beckham
SENIOR PRODUCTION EDITOR   Elizabeth Swain
DESIGNER   Kevin Murphy
ILLUSTRATION COORDINATOR   Edward Starr

This book was set in Palatino by York Graphics and printed and
bound by Courier / Westford. The cover was printed by Phoenix.

This book is printed on acid-free paper.

The paper in this book was manufactured by a mill whose forest management
programs include sustained yield harvesting of its timberlands. Sustained yield
harvesting principles ensure that the numbers of trees cut each year does not exceed
the amount of new growth.

Library of Congress Cataloging-in-Publication Data
RealTime physics : active learning laboratories / David R. Sokoloff,
    Ronald K. Thornton, Priscilla W. Laws.
        p.   cm.
    Contents: Module 1. Mechanics — Module 2. Heat and
thermodynamics.
    ISBN 0-471-12965-8 (set). — ISBN 0-471-28379-7 (module 1). —
ISBN 0-471-28378-9 (module 2)
    1. Physics—Study and teaching—Data processing.   2. Active
learning.   3. Physical laboratories—Study and teaching—Data
processing.   4. Microcomputers.   I. Sokoloff, David R.
II. Thornton, Ronald K.   III. Laws, Priscilla W.
QC30.R344   1998
530′.078′5—dc21   98-5857
                                                        CIP

Printed in the United States of America

10 9 8 7 6 5 4 3

Preface

Development of the series of *RealTime Physics (RTP)* laboratory guides began in 1992 as part of an ongoing effort to create high-quality curricular materials, computer tools, and apparatus for introductory physics teaching.[1] The *RTP* series is part of a suite of *Activity-Based Physics* curricular materials that include the *Tools for Scientific Thinking* laboratory modules,[2] the *Workshop Physics* Activity Guide,[3,4] and the *Interactive Lecture Demonstration* series.[5] The development of all of these curricular materials has been guided by the outcomes of physics education research. This research has led us to believe that students can learn vital physics concepts and investigative skills more effectively through guided activities that are enhanced by the use of powerful microcomputer-based laboratory (MBL) tools.

In the past twelve years new MBL tools—originally developed at Technical Education Research Centers (TERC) and at the Center for Science and Mathematics Teaching, Tufts University—have become increasingly popular for the real-time collection, display, and analysis of data in the introductory laboratory. MBL tools consist of electronic sensors, a microcomputer interface, and software for data collection and analysis. Sensors are now available for motion, force, sound, magnetic field, current, voltage, temperature, pressure, rotary motion, acceleration, humidity, light intensity, pH, and dissolved oxygen.

MBL tools provide a powerful way for students to learn physics concepts. For example, students who walk in front of an ultrasonic motion sensor while the software displays position, velocity, or acceleration in real time more easily discover and understand motion concepts. They can see a cooling curve displayed instantly when a temperature sensor is plunged into ice water, or they can sing into a microphone and see a pressure vs. time plot of sound intensity.

MBL data can also be analyzed quantitatively. Students can obtain basic statistics for all or a selected subset of the collected data, and then either fit or model the data with an analytic function. They can also integrate, differentiate, or display a fast Fourier transform of data. Software features enable students to generate and display *calculated quantities* from collected data in real time. For example, since mechanical energy depends on mass, position, and velocity, the time variation of potential and kinetic energy of an object can be displayed graphically in real time. The user just needs to enter the mass of the object and the appropriate energy equations ahead of time.

The use of MBL tools for both conceptual and quantitative activities, when coupled with recent developments in physics education research, has led us to expand our view of how the introductory physics laboratory can be redesigned to help students learn physics more effectively.

COMMON ELEMENTS IN THE *REALTIME PHYSICS* SERIES

Each laboratory guide includes activities for use in a series of related laboratory sessions that span an entire quarter or semester. Lab activities and homework assignments are integrated so that they depend on learning that has occurred during the previous lab session and also prepare students for activities in the next session. The major goals of the *RealTime Physics* project are: (1) to help students acquire an understanding of a set of related physics concepts; (2) to provide students with direct experience of the physical world by using MBL tools for real-time data collection, display, and analysis; (3) to enhance traditional laboratory skills; and (4) to reinforce topics covered in lectures and readings using a combination of conceptual activities and quantitative experiments.

To achieve these goals we have used the following design principles for each module based on educational research.

- The materials for the weekly laboratory sessions are sequenced to provide students with a coherent observational basis for understanding a single topic area in one semester or quarter of laboratory sessions.

- The laboratory activities invite students to construct their own models of physical phenomena based on observations and experiments.

- The activities are designed to help students modify common preconceptions about physical phenomena that make it difficult for them to understand essential physics principles.

- The activities are designed to work best when performed in collaborative groups of 2 to 4 students.

- MBL tools are used by students to collect and graph data in real time so they can test their predictions immediately.

- A learning cycle is incorporated into each set of related activities that consists of prediction, observation, comparison, analysis, and quantitative experimentation.

- Opportunities are provided for class discussion of student ideas and findings.

- Each laboratory includes a pre-lab warm-up assignment, and a post-lab homework assignment that reinforces critical physics concepts and investigative skills.

The core activities for each laboratory session are designed to be completed in two hours. Extensions have been developed to provide more in-depth coverage when longer lab periods are available. The materials in each laboratory guide are comprehensive enough that students can use them effectively even in settings where instructors and teaching assistants have minimal experience with the curricular materials.

The curriculum has been designed for distribution in electronic format. This allows instructors to make local modifications and reprint those portions of the materials that are suitable for their programs. The *Activity-Based Physics* curricular materials can be combined in various ways to meet the needs of students and instructors in different learning environments. The *RealTime Physics* laboratory guides are designed as the basis for a complete introductory physics laboratory program at colleges and universities. But they can also be used as the central component of a high school physics course. In a setting where formal lectures are given, we recommend that the *RTP* laboratories be used in conjunction with *Interactive Lecture Demonstrations*.

THE HEAT AND THERMODYNAMICS LABORATORY GUIDE

The primary goal of this *RealTime Physics Heat and Thermodynamics* guide is to help students achieve a solid understanding of concepts related to heat and temperature, pressure, the ideal gas law, the laws of thermodynamics, and heat engines. *RealTime Physics Heat and Thermodynamics* includes 6 labs.

Lab 1 (Introduction to Heat and Temperature): This lab is designed to help students distinguish between the concepts of heat and temperature, and to understand the meaning of thermal equilibrium. Temperature sensors and a heat pulser are used to examine the common temperature scales, observe a cooling curve for water, observe two systems come to thermal equilibrium with each other, and examine heat transfer situations in which knowledge of the temperature alone is not sufficient to explain what occurs.

Lab 2 (Energy Transfer and Temperature Change): This lab presents evidence that heat is a form of energy. Semi-quantitative experiments raising the temperature of a system by mechanical and electrical work are followed by a quantitative examination, using a temperature sensor and a heat pulser, of the relationship between the heat energy transferred to a system, the mass of the system, and the change in temperature. Specific heat and the mechanical equilvalent of heat are introduced.

Lab 3 (Heat Energy Transfer): Students use a temperature sensor and a heat pulser to examine the relationship between heat transfer and temperature difference for a sample of hot water in an uninsulated cup. Then various means of decreasing conduction and convection of heat are explored. Finally, the transfer of heat energy by emission and absorption of radiation is explored.

Lab 4 (The First Law of Thermodynamics): Phase changes (ice to water and water to steam) are studied as examples of processes in which the internal energy of a system is increased by a transfer of heat energy. The latent heats of fusion and vaporization are introduced. Then students explore the importance of pressure in describing a gas, the calculation of thermodynamic work, and the first law of thermodynamics. Temperature and pressure sensors are used.

Lab 5 (The Ideal Gas Law): Pressure and tempeature sensors are used to explore Boyle's, Gay-Lussac's, and Charles' laws, leading to the ideal gas law. Kinetic theory is explored at a very basic level through the use of digitized simulation movies.

Lab 6 (Heat Engines): The concept of a cyclic process is introduced through the example of a rubber band engine. Adiabatic and isothermal compressions are introduced with a fire syringe and a small plastic syringe. Finally, a real heat engine consisting of a low-friction glass syringe and hot and cold water heat reservoirs is examined quantitatively using temperature and pressure sensors.

ON-LINE TEACHERS' GUIDE

The *Teachers' Guide* for *RealTime Physics Heat and Thermodynamics* is available online at **http:/www.wiley.com/college/sokoloff-physics.** This *Guide* focuses on pedagogical (teaching and learning) aspects of using the curriculum, as well as computer-based and other equipment. The *Guide* is offered as an aid to busy physics educators and does not pretend to delineate the "right" way to use the *RealTime Physics Heat and Thermodynamics* curriculum and certainly not the MBL tools. There are many right ways. The *Guide* does, however, explain the educational philosophy that influenced the design of the curriculum and tools and suggests effective teaching methods. Most of the suggestions have come from the college, university, and high school teachers who have participated in field testing of the curriculum.

The *On-line Teachers' Guide* has seven sections. Section I presents suggestions regarding computer hardware and software to aid in the implementation of this activity-based MBL curriculum. Sections II through VII present information about the six different labs. Included in each of these is information about the specific equipment and materials needed, tips on how to optimize student learning, answers to questions in the labs and complete answers to the homework.

EXPERIMENT CONFIGURATION FILES

Experiment configuration files are used to set up the appropriate software features to go with the activities in these labs. You will need the set of files which is designed for the software package you are using, or you will need to set up the files yourself. At this writing, experiment configuration files for *RealTime Physics Heat and Thermodynamics* are available with the Vernier software packages *MacTemp* (*Temperature* for MS DOS), *Data Logger* (for Macintosh and MS-DOS) and *Logger Pro* (for Windows and Macintosh), and for the PASCO *Science Workshop* (for Macintosh and Windows). Appendix A of this module outlines the features of the experiment configuration files for *RealTime Physics Heat and Thermodynamics*. For more information, consult the *On-line Teachers' Guide*.

CONCLUSIONS

RealTime Physics Heat and Thermodynamics has been used in a variety of educational settings. Many university, college, and high school faculty who have used this curriculum have reported improvements in student understanding. Their comments are supported by our careful analysis of pre- and post-test data using conceptual evaluation tests. Similar research on the effectiveness of *RealTime Physics Electric Circuits*[6] and *Mechanics*,[7,8,9] also shows dramatic conceptual learning gains in these topic areas. We feel that by combining the outcomes of physics educational research with microcomputer-based tools, the laboratory can be a place where students acquire both a mastery of difficult physics concepts and vital laboratory skills.

ACKNOWLEDGMENTS

RealTime Physics Heat and Thermodynamics could not have been developed without the hardware and software development work of Stephen Beardslee, Lars Travers, Ronald Budworth, and David Vernier. We are indebted to numerous college, university, and high school physics teachers, and especially to Curtis Hieggelke (Joliet Junior College), John Garrett (Sheldon High School), and Maxine Willis (Gettysburg High School) for beta testing earlier versions of the laboratories with their students. At the University of Oregon, we especially thank Dean Livelybrooks for supervising the introductory physics laboratory, for providing invaluable feedback, and for writing some of the homework solutions for the *Teachers' Guide*. Frank Womack, Dan DePonte, and all of the introductory physics laboratory teaching assistants provided valuable assistance and input. We also thank the faculty at the University of Oregon (especially Stan Micklavzina), Tufts University, and Dickinson College for their input into *Tools for Scientific Thinking Heat and Tempeature* and *Workshop Physics*, on which parts of *RealTime Physics Heat and Thermodynamics* are based, and for assisting with our conceptual learning assessments. Finally, we could not have even started this project if not for our students' active participation in these endeavors.

This work was supported in part by the National Science Foundation under grant number DUE-9455561, *"Activity Based Physics: Curricula, Computer Tools, and Apparatus for Introductory Physics Courses,"* grant number USE-9150589, *"Student Oriented Science,"* grant number DUE-9451287, *"RealTime Physics II: Active University Laboratories Based on Workshop Physics and Tools for Scientific Thinking,"* grant number USE-9153725, *"The Workshop Physics Laboratory Featuring Tools for Scientific Thinking,"* and grant number TPE-8751481, *"Tools for Scientific Thinking: MBL for Teaching Science Teachers,"* and by the Fund for Improvement of Post-secondary Education (FIPSE) of the U.S. Department of Education under grant number G008642149, *"Tools for Scientific Thinking,"* and number P116B90692, *"Interactive Physics."*

REFERENCES

1. Ronald K. Thornton and David R. Sokoloff, "RealTime Physics: Active Learning Laboratory," in *The Changing Role of the Physics Department in Modern Universities, Proceedings of the International Conference on Undergraduate Physics Education,* 1101–1118 (American Institute of Physics, 1997).

2. Ronald K. Thornton and David R. Sokoloff, "Tools for Scientific Thinking—Heat and Temperature Curriculum and Teachers' Guide," (Portland, Vernier Software, 1993) and David R. Sokoloff and Ronald K. Thornton, "Tools for Scientific Thinking—Motion and Force Curriculum and Teachers' Guide," Second edition, (Portland, Vernier Software, 1992).

3. P. W. Laws, "Calculus-based Physics Without Lectures," *Phys. Today* **44**: 12, 24–31 (December, 1991).

4. Priscilla W. Laws, *Workshop Physics Activity Guide: The Core Volume with Module 1: Mechanics,* (Wiley, New York, 1997).

5. David R. Sokoloff and Ronald K. Thornton, "Using Interactive Lecture Demonstrations to Create an Active Learning Environment," *The Physics Teacher* **27**: 6, 340 (1997).

6. David R. Sokoloff, "Teaching Electric Circuit Concepts Using Microcomputer-Based Current and Voltage Probes," chapter in *Microcomputer-Based Labs: Educational Research and Standards,* Robert F. Tinker, ed., *Series F, Computer and Systems Sciences,* **156**, 129–146 (Berlin, Heidelberg, Springer Verlag, 1996).

7. Ronald K. Thornton and David R. Sokoloff, "Assessing Student Learning of Newton's Laws: The *Force and Motion Conceptual Evaluation* and the Evaluation of Active Learning Laboratory and Lecture Curricula," *Am. J. Phys.* **64**, 338 (1998).

8. Ronald K. Thornton, "Learning Physics Concepts in the Introductory Course: Microcomputer-based Labs and Interactive Lecture Demonstrations," in *Conference on the Introductory Physics Course,* J.W. Wilson, ed. (Wiley, New York, 1997), pp. 69–85.

9. Ronald K. Thornton and David R. Sokoloff, "Learning Motion Concepts Using Real-Time Microcomputer-Based Laboratory Tools," *Am. J. Phys.* **58**, 858 (1990).

This project was supported, in part, by the Fund for the Improvement of Post-Secondary Education (FIPSE) and the National Science Foundation. Opinions expressed are those of the authors and not necessarily those of the foundations.

Contents

Name _____   Date _____

Pre-Lab Preparation Sheet for Lab 1:
Introduction to Heat and Temperature
(Due at the beginning of Lab 1)

Directions:
Read over Lab 1 and then answer the following questions about the procedures.

1. What are the approximate Fahrenheit temperatures asked for in Table 1-1?

2. Sketch your Prediction 2-3 on the axes below.

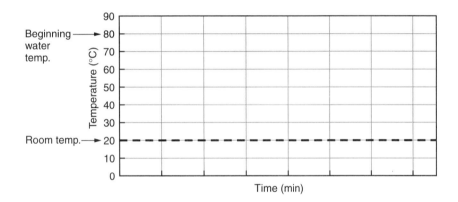

3. What will the two temperature probes be used for in Activity 2-1?

4. What is your Prediction 2-5?

5. What is the heat pulser used for in Activity 3-1?

LAB 1:
INTRODUCTION TO HEAT AND TEMPERATURE

. . . . thermometer readings alone do not tell the entire story of thermal interactions, . . . something else must be happening, and . . . an additional concept (or concepts) must be invented.

— Arnold Arons

OBJECTIVES

- To acquire an operational definition of temperature and to understand the connection between temperature and thermal equilibrium.

- To understand how the temperature of a body changes as it undergoes thermal interactions with its surroundings.

- To understand the difference between the quantity of heat energy transferred to or from an object and its temperature.

OVERVIEW

You are sitting in a hot bathtub with a cup of hot coffee on the side. The coffee and the bath water are at the same temperature. Both are "hotter" than the air in the room. However, it costs much less to heat the coffee than it did to heat the bath water. Why? Can the words heat and temperature be used the same way? Is there any evidence that heat is a substance? In this lab you will learn how contemporary physicists define and use familiar terms like "heat" and "temperature" in a way that helps them to understand thermal processes.

With this lab we begin the study of *thermodynamics,* a way of looking at physical phenomena that is very different from studying mechanics. Thermodynamics is the study of heat transfer and the resulting temperature changes. Much of the physics studied in mechanics involves motions that we can see, while many of the changes we will encounter in thermodynamics will *not* be visible without the help of indirect measuring instruments such as temperature sensors and pressure sensors.

While we will use some of the concepts from mechanics, such as work and kinetic energy, in our discussion of thermodynamics, we will also introduce some new terms. Although you have already encountered some of these terms, such as *heat* and *temperature* in everyday situations, we will need to define them more precisely. Other new terms such as *adiabatic* and *isothermal,* which will be introduced in future labs, are probably less familiar but also very useful.

Temperature is one of the most familiar and fundamental thermodynamic quantities and it is the major focus of study in this first lab. In the first investigation of this lab, you will look at how temperature can be measured using a glass bulb thermometer and using an electronic temperature sensor interfaced with a microcomputer. You will use these to explore the Celsius and Kelvin temperature scales.

In later activities you will take a much more careful look at the concept of temperature. In particular, you will observe how the temperature of a substance or system is affected when it interacts with its surroundings or another substance at a different temperature. Whenever the temperature of something changes, we say glibly that it has undergone a *thermal interaction*. When the temperature of a system remains constant we refer to it as being in *thermal equilibrium*. Since we cannot *see* what really goes on when something changes temperature, we have to develop some new concepts to try to explain what is happening. One of these new concepts is that of *heat transfer*. It is essential to understand the difference between the temperature of an object and the heat transferred to or from the object. This will be a major focus of this lab.

INVESTIGATION 1: TEMPERATURE MEASUREMENT

Temperature is a familiar concept to all of us. Thermometers register changes in temperature by using materials that change in some way as they are heated or cooled. For example, the column of liquid (alcohol or mercury) in a common thermometer expands when heated and contracts when cooled. Thus, the length of the column is longer or shorter depending on its temperature, and the instrument can be used to measure temperature. Based on this type of thermometer, we could crudely define temperature as *a quantity that is related to the height of a column of liquid inside a familiar glass bulb thermometer.*

If the length of the column is associated with standard temperature units and scales such as Fahrenheit or Celsius, the thermometer is said to be *calibrated*. You are probably familiar with the Fahrenheit scale from weather forecasts and you may have worked with the Celsius scale in other science courses or encountered it in other countries. These are just two of the many temperature scales set up by early investigators of thermodynamics. Most of these scales were established by taking two "fixed points" that were reliable, easily reproducible temperatures associated with known physical situations. The investigator decided arbitrarily how many "degrees" lay between these two temperatures.

The Fahrenheit scale was set up by Herr Fahrenheit in Germany in 1724. The zero point (0°F) of his scale was supposed to be the lowest temperature attainable with a mixture of ice and salt, while the upper point was human body temperature, which he called 96°F. On his original scale the freezing point of water exposed to air at sea level turned out to be about 32°F and the boiling point of water exposed to air at sea level turned out to be about 212°F.

In 1742 a Swedish investigator named Celsius devised another scale, which he referred to as the centigrade scale. On this scale the freezing point of water exposed to air at sea level was fixed at 0°C and the boiling point of water exposed to air at sea level was fixed at 100°C. The modern Celsius scale is based on a degree that is the same "size" as the centigrade degree, but is fixed so that the *triple point* of water (i.e., the temperature and pressure at which solid ice, liquid water, and water vapor can coexist) is 0.01°C.

There is one other temperature scale that is considered to be the fundamental scale in thermodynamics: the Kelvin temperature scale, named for Lord Kelvin. The degrees on the Kelvin scale, indicated with a K (rather than °K) are the same size as Celsius degrees. However, the triple point of water is 273.16 K. Therefore $T(K) = T(°C) + 273.15$ K. Zero kelvins $(-273.15°C)$ is called absolute zero, and is believed to be the lowest possible temperature.

When we study electrical phenomena, you will discover that the ease with which electric currents can be carried by certain materials changes with temperature. Thus, it is also possible to use electronic devices to measure temperature. Such systems have several advantages over glass bulb thermometers. The sensors can be much smaller and usually respond more quickly to changes in temperature. Using a computer, you can automatically produce graphs of temperature vs. time for one or several sensors at a time. And, as usual, the data you collect can be displayed in tabular form and analyzed.

In the first activity of this lab you will use both a familiar glass bulb thermometer and an electronic sensor to measure the temperatures of objects around the room. The purpose of this activity is to become familiar with the Celsius and Kelvin temperature scales. You will also gain some experience with electronic temperature measurement, some of the limitations of electronic sensing, and features of the temperature measuring software.

To conduct this activity you will need

- glass thermometer, calibrated with Celsius scale (-5 to 105°C)
- computer-based laboratory system
- 2 temperature sensors
- temperature software
- *RealTime Physics Heat and Thermodynamics* experiment configuration files
- hot water (about 80°C)
- crushed ice and cold water
- Styrofoam cup, 300 mL, (for crushed ice)
- vat (to prevent spills)

Activity 1-1: Glass Bulb and Electronic Thermometers

1. To get started with the electronic temperature sensors, first be sure that they are plugged into the appropriate connectors on the interface. Open the experiment file called **Digital Readouts (L1A1-1)** so that the two temperature sensors are displayed in degrees Celsius.

2. Use the **calibrate** feature of the software to **load a calibration file** for the two sensors.

3. Use the two temperature sensors and the Celsius glass bulb thermometer available to you to measure the temperatures of the air, body temperature (inside elbow), ice water, room-temperature water, and boiling water. Be sure that the three thermometer tips are sensing the same temperature by placing them right next to each other. Stir the water and ice samples to be sure that the whole sample is at the same temperature.

   Record the temperatures in Table 1-1.

4. Fill in the approximate corresponding Fahrenheit temperatures that you know.

Table 1-1

	Glass bulb thermometer temperature (°C)	Electronic temperature sensors (°C)		Approximate Fahrenheit temperature (°F)	Calculated Kelvin temperature (K)
		Sensor 1	Sensor 2		
Room air					
Body (inside elbow)					
Ice water (equal amts. of ice and water, stirred)					
Room temperature water					
Boiling water					

Question 1-1: Do the readings of the three thermometers seem to agree? If not, why might they not agree? (**Hint:** See the **Comment** below.)

Question 1-2: Calculate the Kelvin temperatures corresponding to the glass bulb thermometer temperatures in Table 1-1 and record them in the last column of the table.

Comment: An electronic temperature sensor used in combination with a particular computer and interface must be calibrated against a glass bulb thermometer or some other known standard. This can be accomplished by using the **calibrate** feature of your temperature software.

---

5. If the readings of your electronic temperature sensors differ from each other or from the glass bulb thermometer by more than ±0.5°C, then you need to recalibrate the electronic sensors.

6. Find the **calibrate** feature in your temperature software and follow the instructions given on the computer screen. Put both temperature sensors simultaneously into the same samples of water at known temperatures.

   Ice water and hot tap water will work fine for the two different temperatures needed for calibration. Use your marked glass bulb thermometer as a standard.

   Since you will probably be using the same temperature sensors and interface again during other labs, you should **name and save your calibration file.**

7. If you recalibrate the sensors, repeat the measurements above and correct the temperature readings in the table.

Question 1-3: Any of the temperatures you have measured might be used as *fixed points* to define a temperature scale. Which of these might be reliable ones; i.e., which ones are truly *repeatable temperatures?* Explain.

Question 1-4: What is the number of divisions between the freezing point and boiling point of water on the Celsius, Kelvin, and Fahrenheit scales? Which degree is larger, the °C, K, or °F?

Prediction 1-1: When a nurse pops a room temperature glass bulb thermometer in your mouth to see if you have a fever, he leaves the thermometer in your mouth awhile before reading the temperature. Why does he wait?

Prediction 1-2: Suppose you want to measure room temperature with a thermometer that has been in ice water. If you want an accurate value, for which would you need to wait longer: measuring room temperature water or room temperature air? Explain the reason for your prediction.

Use your electronic temperature sensors to test your predictions.

Activity 1-2: Reaching Thermal Equilibrium

1. Open the experiment file called **Reaching Equilibrium (L1A1-2)** to set up your software to graph two temperature sensors vs. time over a range of 0 to 30°C for a time interval of 240 s.

2. **Load the calibration file** for the two temperature sensors that you saved in Activity 1-1, if it hasn't already been loaded.

3. With the container of room-temperature water nearby, start with both sensors in ice water. **Begin graphing,** and at the same time move one sensor into the air and the other into room-temperature water.

4. Record the time intervals for the temperature sensors to reach room temperature.

Ice water to room air temperature: $\Delta t = $ _____ s

Ice water to room temperature water: $\Delta t = $ _____ s

Question 1-5: Which reached room temperature faster? On the basis of these measurements what should you watch out for in making temperature measurements?

Prediction 1-3: Suppose you take two thermometers out of hot water, dry one, and then wave both around in the air. Will there be any difference in the time it takes them to reach room temperature? (**Hint:** Remember how it feels to get out of a shower on a dry day? Brrr!)

5. **Adjust the temperature axis** so that it includes the temperature of your hot water.

6. **Begin graphing** with both temperature sensors in the hot water. At the same moment, dry just one of the probes, and begin shaking both vigorously in the air.

7. Record the time intervals for the temperature sensors to reach room temperature.

Hot water to room air (dried): $\Delta t =$ _____ s

Hot water to room air (wet): $\Delta t =$ _____ s

Question 1-6: Which reached room temperature faster? Can you explain why? On the basis of these measurements, what should you watch out for in making temperature measurements?

Comment: A temperature sensor always measures *its own temperature*. When you touch the sensor to something, it takes a few seconds for the sensor and the object to reach a common temperature. After this happens the sensor is said to be in *thermal equilibrium* with the object it is touching. To make accurate temperature measurements it is important to wait until the temperature reading of the sensor remains fairly constant (until the temperature sensor and the object are in thermal equilibrium). As you have just seen, for a given thermometer this time lag varies considerably, depending on what system's temperature you are measuring. You also need to beware of *cooling by evaporation*. Be careful not to measure air temperatures when the sensitive part of the thermometer is wet (especially with alcohol). Evaporation of liquid from the tip of the thermometer will cool it.

INVESTIGATION 2: THERMAL EQUILIBRIUM AND HEAT TRANSFER

You have already seen that hot objects cool down until they reach the same temperature as objects around them. This process is also described as reaching *ther-*

mal equilibrium. In this investigation you will examine *how* a hot liquid, say a cup of hot chocolate on a table (in a room like this one), cools down. You will also examine what happens to the temperatures of objects nearby when something cools down.

Observations such as these lead most people to the conclusion that objects at different temperatures interact with each other through the transfer of something from hotter objects to cooler ones.

Let's begin with some predictions. Suppose you have a cup of hot chocolate. You place it on the lab table and let it sit for a long while.

Prediction 2-1: How cool will the hot chocolate get?

Prediction 2-2: Where does the heat go as the hot chocolate cools?

Prediction 2-3: Sketch on the axes below your prediction of how the temperature of the hot chocolate will change with time. (Assume that the beginning temperature is 80°C and that room temperature is 20°C.)

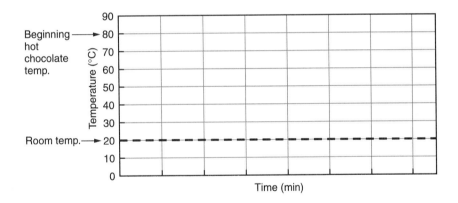

Prediction 2-4: Is the *rate* at which the temperature of the hot chocolate drops always the same or does it change as the hot chocolate gets cooler? If it changes, is the rate faster when the hot chocolate is hotter or cooler?

To test your predictions, you will need

- computer-based laboratory system with 2 temperature sensors
- temperature software
- *RealTime Physics Heat and Thermodynamics* experiment configuration files
- hot water (at least 80°C) or immersion heater to heat water

- container marked in mL

- *uninsulated* glass, plastic, metal, or paper cup

- vat (to prevent spills)

Activity 2-1: Cooling Water—A Temperature History

1. Open the experiment file called **Cooling Water (L1A2-1).** This will set up the temperature software to graph temperature sensor 1 vs. time for 10 min on axes similar to those shown below and display both temperature sensors digitally.

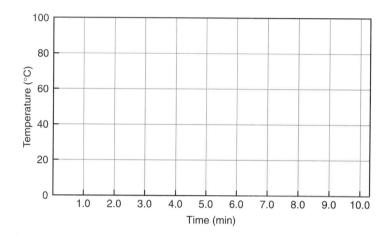

2. **Load the calibration file** for the two temperature sensors that you saved in Activity 1-1, if it hasn't already been loaded.

3. Measure room (air) temperature with sensor 1 and record it in column 1 of Table 2-1 after its reading remains constant.

4. Put temperature sensor 1 in the *uninsulated* cup. Add a small amount of hot water (about 50 mL at 80°C or higher) and immediately **begin graphing.** *Do step 5 immediately.*

5. See if you can find where the heat is transferred as the hot water cools by using sensor 2 to see what is getting warmer. Sensor 1 will continue to measure the water temperature.

   Fill in the temperatures in Table 2-1 using sensor 2—once near the beginning of the experiment, then again after about 5 min. *For each temperature reading, be sure to wait long enough for the sensor to reach the temperature you are measuring.* Fill in all of column 1, then do column 2.

6. Record the temperature of the water after 10 min.

7. Leave the cup of water for the rest of the lab and measure its temperature and room (air) temperature again just before you leave. Record these in the table.

8. **Print your graph** and affix it over the previous axes.

9. Calculate the temperature change of the water in the first 5 min and in the last 5 min.

Δ temperature in first 5 min: _____ °C

Δ temperature in last 5 minutes: _____ °C

Table 2-1

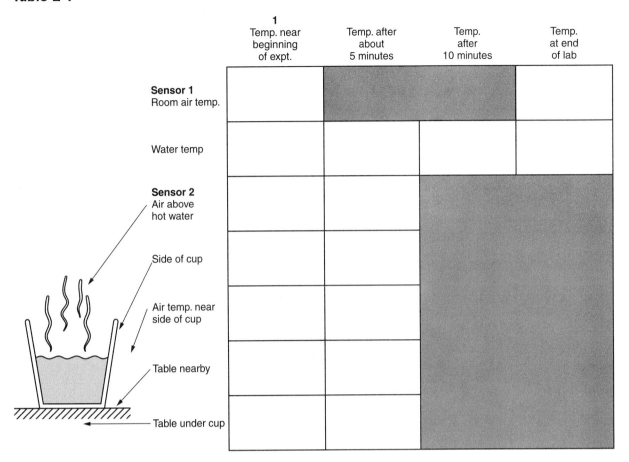

	1 Temp. near beginning of expt.	Temp. after about 5 minutes	Temp. after 10 minutes	Temp. at end of lab
Sensor 1 Room air temp.				
Water temp				
Sensor 2 Air above hot water				
Side of cup				
Air temp. near side of cup				
Table nearby				
Table under cup				

Question 2-1: As the hot water cools, has the temperature of anything else increased (relative to room temperature)? What things had the largest temperature rise according to your measurements?

Question 2-2: Does any matter flow through the walls or bottom of the cup to its surroundings? (Does any water actually leave the cup?) Is a change in temperature always associated with a flow of matter?

Question 2-3: Did the temperature drop more in the first 5 min or in the last 5 min? Why do you think this happened? What do you conclude about the rate of cooling of water as the temperature of the water drops?

We know that when a hotter substance comes into thermal contact with a cooler one the temperatures of the two substances change. This temperature change is easy to observe in situations where substances (e.g., liquids) can be mixed together or come into thermal contact with each other without mixing. Do the initial temperatures alone allow us to predict the final temperature of the system after the two substances have interacted with each other? In the next activity you will examine more carefully this interaction between hot and cold objects. First make a couple of predictions.

Prediction 2-5: Suppose you have a mass of hot water at 80°C in a small, sealed, uninsulated container and a larger mass of water at room temperature (20°C) in a cup. The container is submerged in the water in the cup and left there for a long time.

A. What do you predict will be the final temperature of the water in the container? Will it be midway between 80 and 20°C? Or will it be closer to 80°C or closer to 20°C?

B. What do you predict will be the final temperature of the water in the cup? Will it be midway between 80 and 20°C? Or will it be closer to 80°C or closer to 20°C? Why?

C. Is it possible knowing just the initial temperatures of the water (and not the masses) to predict *exactly* what the final temperature will be? Explain.

To test your predictions you will need the following materials in addition to those you used in Activity 2-1:

- more hot water (about 80°C)
- 35-mm-film container with a hole in the cover just large enough for a temperature sensor
- Styrofoam or other *insulated* cup

Activity 2-2: Temperature Change and Heat Transfer

1. Use the film container to pour enough full containers of room-temperature water into the cup so that the film container can be submerged in the water. Put temperature sensor 1 in the cup.

    Number of containers full of water put in the cup: _____

2. Carefully push temperature sensor 2 through the hole in the film container cover. It should be a tight fit.

    [When you are done with this activity, remove the probe from the cover very gently by pushing (not pulling) it through the hole.]

3. Open the experiment file called **Temp. Change and Heat Transfer (L1A2-2)** to display and graph both sensors on axes like those that follow.

4. Fill the film container with water near 80°C, and seal it with the cover and sensor.

5. Quickly do the following. Record the initial temperatures.

Hot container: _____ Cool cup: _____

Begin graphing, and then submerge the film container in the cup. Gently move the container up and down to mix the water in the cup for the rest of this activity.

6. When you are done, **print your graph** and affix it over the axes.

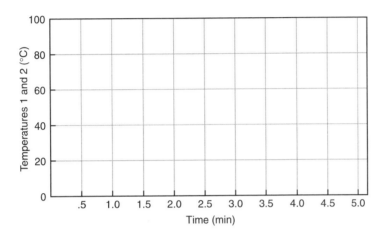

Question 2-4: Did the final temperatures of the hot water and cool water agree with your prediction? Does the final temperature seem to depend on anything besides the initial temperatures of the water in the container and in the cup? Explain. (**Hint:** Compare the masses of hot and cold water and the *change in temperature* of each.)

What would happen to the final temperature if you changed the relative amounts of water so that there was the same amount of hot water in the film container but less cool water in the cup?

Question 2-5: As the hot water cooled down, what happened to the room-temperature water? Is it likely that an exchange of water between the container and the cup accounts for the temperature changes? If not, how did the two temperatures change even though no hot water mixed with the room-temperature water?

Question 2-6: How can you tell when thermal equilibrium has been reached? Is there any evidence that whatever flowed from one container of water to the other has stopped flowing? What is this evidence?

Comment: As you have observed in this investigation, inevitably, parts of any thermally isolated (i.e., insulated) system having different temperatures will interact until the entire system is at the same temperature. This is a mysterious process, because the interaction that causes temperature changes in two parts of a system until thermal equilibrium is reached *can occur without an exchange of matter.*

You should have noticed from your experiments that the relative masses of the parts of your thermally isolated system affect the value of the final equilibrium temperature. Thus, the interaction between two parts of a system cannot be explained as a simple *temperature* exchange. We need to create a new concept to help us understand heating and cooling processes. Scientists have invented the term *heat transfer* to explain this phenomenon.

The use of the noun "heat" is misleading, since using this term to explain temperature changes implies the exchange of a substance between two parts of a system. The word "heat" is actually a sloppy shorthand for an interaction process that leads to temperature changes. As a reminder that we are dealing with a process rather than a substance, we will use the term *heat transfer* and not simply *heat*. In the next investigation, you will further explore the relationship between heat transfer and temperature.

We will explore more carefully in the next lab why heat is actually considered to be a form of energy. Thus what people casually refer to as "heating" is actually what scientists refer to as *heat energy transfer.*

INVESTIGATION 3: TEMPERATURE IS NOT THE WHOLE STORY

You know that one way to raise the temperature of a substance like water is to "heat" it, say on a stove or with a small immersion heating coil. We will explore more carefully in the next lab how heat is transferred by an electric heating coil. For now, we will simply use an *electric heat pulser* as a research tool to explore further the relationship between *heat transfer* and *temperature*.

Your computer-based temperature system can be outfitted with a heat pulser consisting of an immersion heater connected to the computer interface. Each time you push the **HEAT** button or key, a pulse of the same amount of heat is transferred to the system you are examining.

You will first explore a situation where heat is transferred to water and yet *the temperature of the water does not change.* Then you will examine the amounts of heat transfer needed to bring about the same temperature change in different masses of water. First a prediction:

Prediction 3-1: If you transfer heat to water in a cup at the same rate as it is leaving, what will happen to the temperature of the water? What will happen if you transfer in more heat than is transferred out? Less than is transferred out? Explain.

In the first activity, you are going to try to keep hot water at a constant temperature by transferring heat with the heat pulser to replace the heat that is transferred out to the surroundings. You will need the following:

- computer-based laboratory system with one temperature sensor
- temperature software
- *RealTime Physics Heat and Thermodynamics* experiment configuration files
- heat pulser (relay box and immersion heater)
- hot water (near 80°C)
- stirring rod
- container marked in mL
- *uninsulated* glass, plastic, metal, or paper cup
- vat (to prevent spills)

Activity 3-1: Keeping Hot Water Hot

> **Warning!!** Do not plug the immersion heater into the pulser or press the HEAT button or key unless the heater is in the water. The heater will burn out if it is turned on even for a short time while not immersed in water.

1. Plug the temperature sensor and the heat pulser into the interface.

2. Open the experiment file called **Keeping Hot (L1A3-1)** to set up the axes shown below. This will also set up the heat pulser to transfer 2-s pulses of heat each time the **HEAT** button or key is pressed.

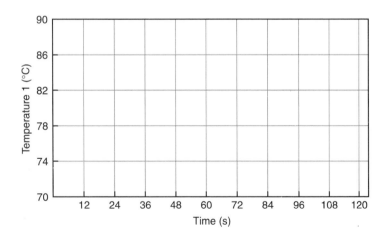

3. **Load the calibration** for the temperature sensor, if this hasn't already been done.

4. Pour about 100 mL of hot water (near 80°C) into the *uninsulated* uncovered cup. Put the sensor and heater into the water, and plug the heater into the relay box just before starting. (Use the heat pulser to warm the water to 80°C before you start, if necessary.)

Water temperature at the start: _____°C

5. **Begin graphing,** and *while stirring continuously* try to keep the water temperature constant (within a degree). Press the **HEAT** button or key when you need to transfer more heat to the water. The computer will make marks on the graph each time you pulse the heater and will keep track of the number of pulses you use. *Keep stirring!*

6. When you are done, **print your graph** and affix it over the previous axes.

7. Record the number of heat pulses transferred: _____

Question 3-1: When you heat water below its boiling point in a tea kettle on the stove, the temperature of the water rises. How is it possible that in this activity heat was transferred to the water in the cup and yet the temperature remained the same?

Question 3-2: In this experiment there was no temperature change. What factor besides temperature change might determine the amount of heat transfer that was needed?

Question 3-3: In this experiment you are working as part of a *feedback loop* to determine when more heat is needed to keep the temperature of the water constant. Can you think of a device in your home that operates in a similar way?

If you have additional time, carry out Extension 3-2 in which you will try to raise the temperature of different masses of water in an insulated cup *by the same amount* by transferring heat from the heat pulser to the water.

Extension 3-2: Heating Up Water

Prediction E3-2: Heat is transferred to water in a perfectly insulated cup at a steady rate for 80 s, and then no more heat is transferred. No heat can leak in or out. Sketch below your prediction for the graph of the temperature of the water as a function of time.

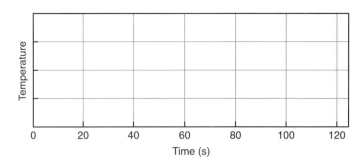

To test your prediction you will need the following materials in addition to those from the last activity:

- Styrofoam or other *insulated* cup

- room-temperature water

REALTIME PHYSICS: HEAT AND THERMODYNAMICS

1. Open the experiment file called **Heating Water (L1E3-2)** to display the axes that follow. This will also set up the heat pulser to transfer 5-s pulses of heat.

2. Put the temperature sensor and the heater into the *foam (insulated)* cup. Add 75 g (mL) of water near room temperature to the cup. Be sure that the water covers the coils of the heater.

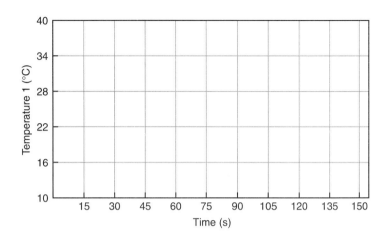

3. Measure the initial temperature of the water and record it in Table E3-2.

4. **Begin graphing.** Stir the water vigorously during the entire time the computer is graphing the temperature.

   For the first 10 s don't pulse the heater. Just watch the temperature when no heat is transferred. Then pulse the heater every 10 s for a total of 4 heat pulses. *Keep stirring!*

5. When the temperature stops increasing, record the final temperature in Table E3-2.

6. Calculate the temperature change of the 75 g (mL) of water and the temperature change per heat pulse, and record these in the table.

7. Use the features of your software to transfer your data so that the graph will remain **persistently displayed on the screen** for comparison during the next activity.

Question E3-4: Why did the temperature of the water rise during this activity, while it remained constant as heat was transferred in the previous activity?

Table E3-2

Mass of water (g)	Initial temperature (°C)	Final temperature (°C)	Temperature change (°C)	Number of heat pulses transferred	Temperature change per pulse (°C/pulse)
75				4	
150					

Question E3-5: Did the shape of the graph agree with your prediction? Describe any differences.

Prediction E3-3: Suppose you started with twice as much water (150 g) at room temperature and wanted to heat the water to the *same final temperature* as you did the 75 g. How many heat pulses would you need to transfer? Four? More than four? Less than four? Explain how you arrived at your prediction.

8. Test your prediction by repeating the procedure above, this time using 150 g of room temperature water, and transferring as many heat pulses as you need to bring about the *same temperature change. Remember to stir vigorously throughout the experiment.*

9. **Print your graph** with both experiments displayed, and affix it over the previous axes.

Question E3-6: Did your results agree with your prediction? Describe any differences.

Question E3-7: Assume that very little heat is exchanged between the insulated cup and its surroundings while you transfer heat pulses to the water in it. How do you explain your observation that different numbers of heat pulses were needed to cause the *same temperature change* in these two experiments?

Question E3-8: In this experiment, what factor in addition to the temperature change affected the amount of heat that was transferred?

Question E3-9: Since heat was transferred by the heat pulser at a steady rate in time, the time and the amount of heat transfer are proportional; i.e., the horizontal axis could be re-labeled as "amount of heat transferred." From your graphs, what type of relationship appears to exist between the temperature of the water and heat transferred? Explain.

Question E3-10: How would the temperature changes for equal heat transfer to different amounts of water depend on the mass of the water? Describe the relationship between the temperature change (ΔT) and the mass (m) for a fixed amount of heat transferred in words and mathematically. (**Hint:** Look at your data for temperature change per pulse and at the slopes of your graphs.)

If you doubled the mass of the water and transferred the same amount of heat, what would happen to the temperature change?

HOMEWORK FOR LAB 1:
INTRODUCTION TO HEAT AND TEMPERATURE

1. Approximately what is human body temperature on the Celsius temperature scale and on the Kelvin scale? What is room temperature?

2. Why must a thermometer be calibrated? You have a glass bulb thermometer with no markings on it and a second one marked in degrees Celsius. Explain how you would calibrate the unmarked thermometer.

3. Compare the length of time it takes the temperature sensor to reach thermal equilibrium with the air or water it is in. Is it easier to exchange heat between a thermometer and air or between a thermometer and water.

4. What effect does the evaporation of water have on the object from which it is evaporating? When has your body experienced this effect? During evaporation, heat is being exchanged. What is it transferred from and what is it transferred to?

5. If you put a cup of hot chocolate at 90°C on a table in a room kept at 25°C you know it will cool down.

    a. How cool will the hot chocolate get?

    b. If you put it outside where the temperature is 5°C, how cool will it get?

c. Compare the initial rates of cooling in (a) and (b). Are they the same or is one larger? Explain.

d. Where does the heat go as the hot chocolate cools down?

e. On the axes below, sketch a graph of the temperature of the hot chocolate vs. time if the hot chocolate starts at 90°C and is placed in a room where the temperature is kept at 25°C.

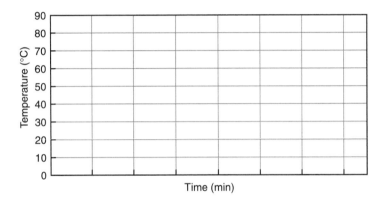

f. Explain the shape of the graph, especially changes in the rate of cooling as the hot chocolate cools down, in terms of your observations in this lab.

6. You have a mass of hot water at 90°C and twice the mass of cool water at 10°C. The hot and cold water are mixed together. Will the final temperature be midway between 10 and 90°C, closer to 90°C, or closer to 10°C? Explain your reasoning.

7. Define *thermal equilibrium*. How can you tell that two objects in thermal contact with each other are in thermal equilibrium?

8. The two containers of water below are completely insulated so that no heat can be transferred in or out. The water in both containers started at room temperature (20°C), and heat was transferred to both using heating coils until they reached the indicated final temperatures. Which container had more heat transferred to it? Explain how it is possible to transfer different amounts of heat to each container and get the same temperature change.

9. The two containers of water below are completely insulated. (No heat can be transferred in or out.) The water in both containers started at room temperature (20°C), and heat was transferred to both using heating coils until they reached the indicated final temperatures. Which container had more heat transferred to it? Explain your answer.

10. A pot of water is heated on a stove. After the temperature of the water has increased a certain amount (but not to the boiling point), the temperature stops rising even though heat continues to be transferred to the water. Explain how it is possible to transfer heat to the water without changing the temperature.

11. Draw with a solid line on the axes that follow the shape of the temperature–time graph that results when you transfer heat at a steady rate to an insulated cup of water that started out at 20°C. Draw with a dashed line on the same axes the shape of the temperature–time graph of the water that results when you put an 80°C solid object in an insulated cup of 20°C water. What is it about heat transfer and its dependence on temperature *difference* that explains the two different shapes of these graphs?

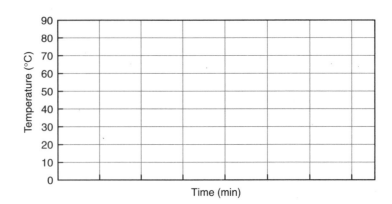

12. Summarize the evidence from this lab that tells you that temperature alone is not enough to describe what is going on when objects at different temperatures are placed in thermal contact with each other. How does heat transfer seem to figure into this? What is the difference between temperature and heat transfer?

Name_____   Date_____

PRE-LAB PREPARATION SHEET FOR LAB 2:
ENERGY TRANSFER AND TEMPERATURE CHANGE
(Due at the beginning of Lab 2)

Directions:
Read over Lab 2 and then answer the following questions about the procedures.

1. What is your Prediction 1-1?

2. What is your Prediction 2-1?

3. Why are first 4 and then 8 pulses of heat transferred to the same mass of water in Activity 2-1?

4. How is specific heat capacity defined? How will you find the specific heat capacity of water in Activity 2-2?

5. What is the meaning of *mechanical equivalent of heat?*

LAB 2:
ENERGY TRANSFER AND TEMPERATURE CHANGE

. . . the quantity of heat produced by the friction of bodies, whether solid or liquid, is always proportional to the quantity of energy expended.

—James Joule

OBJECTIVES

- To establish the concept of *heat* as *heat energy transfer* from a substance at a higher temperature to one at a lower temperature.

- To quantify the relationship between the heat energy transferred to a system and the change in temperature of the system.

- To understand the meaning of *specific heat* and measure its value for several liquids.

- To determine the equivalence between the common unit of heat energy, the *calorie*, and the unit of energy, the *joule*.

OVERVIEW

So far you have made observations that indicate that interactions take place when two substances in thermal contact are at different temperatures. We have called these interactions "heat energy transfer." There are other ways to raise the temperature of an object. For example, you could rub a piece of metal on emery paper or sandpaper and measure its temperature increase with a thermometer. Also, as you have observed in Lab 1, it is possible to produce a temperature increase using an electric heater by supplying electrical energy to it. Observations like these caused physicists and engineers in the middle of the nineteenth century to conclude that heat is just a form of energy, the form that flows when there is a temperature difference between two objects. Today a physicist or engineer would say that *heat is a form of heat (or thermal) energy transfer.*

While you have already observed a relationship between the heat energy transferred to a system and the temperature change of the system, in this lab you will examine the mathematical relationship between these quantities. In doing so, you will look at the amount of heat energy transfer needed to raise the temperature of one unit of mass of a substance by one degree, which is called the *specific heat* of the substance. You will find the specific heat of water and another liquid.

The common unit of heat energy is the calorie. Mechanical energy is measured in joules, a familiar unit used in your study of mechanics. In the final investigation of this lab, you will compare the number of calories of heat energy producing a measured temperature change in water to the number of joules of energy delivered by the heat pulser to find the equivalent of a calorie of heat energy in joules. This number that converts calories into joules is commonly known as the *mechanical equivalent of heat.*

INVESTIGATION 1: HEAT TRANSFER AS ENERGY TRANSFER

In this investigation you will explore ways of raising the temperature of a system by converting other forms of energy into heat energy that is transferred to the system. You will start by observing what happens when you do mechanical work on the system and then when you do electrical work on the system.

Prediction 1-1: If you hold a piece of metal in your hand and rub it back and forth on emery paper or sandpaper, do you expect the temperature of the metal to change? How will it change?

Suppose that you rub the metal back and forth for twice as long a time. Will the temperature change be different from before? If so, how will the temperature change differ?

To test your predictions you will need

- computer-based laboratory system
- temperature sensor
- temperature software
- *RealTime Physics Heat and Thermodynamics* experiment configuration files
- fine emery paper or sandpaper
- iron nail
- piece of metal with embedded electrical heater, hole for temperature sensor, and foam insulation with which to hold it

Activity 1-1: Mechanical Work and Temperature Change

1. Grasp the nail near its head with your fingers, with the head facing down, and rub it vigorously back and forth on the emery paper.

Question 1-1: Describe what you felt. Was there a temperature change?

Question 1-2: Was there any transfer of matter to the nail as the rubbing took place? Was there any evidence of mechanical work being done while the rubbing took place? Explain.

2. Set up the temperature sensor in the hole in the piece of metal. Use a piece of tape to hold it in place. Make sure that the wires are out of the way so that you can rub the opposite side on the emery paper.

3. Open the experiment file called **Mech. and Elect. Work (L2A1-1)** to plot one temperature sensor on the axes shown below.

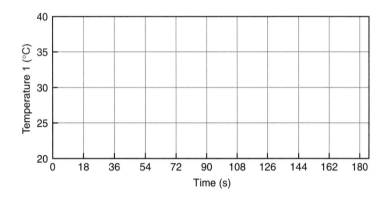

4. **Load the calibration file** for your temperature sensor.

5. **Begin graphing,** and then after 10 s, hold the metal using the foam insulation and begin rubbing the bottom of the metal on the emery paper. Rub for 20 s, applying a uniform force and moving the metal back and forth over a distance of about 10 cm (3 inches) on the emery paper.

6. After the 20 s of rubbing, hold the metal above the emery paper (grasping it with the foam) until the temperature stops changing. Then wait about 5 s more, *note the time,* and rub for 40 s more. Try to rub with the same force and over the same distance as before.

   Again hold the metal with the foam, this time until the end of the 180 s.

7. If you are going to do Extension 1-2, then transfer your data so that the graph will **remain persistently displayed on the screen** for later comparison.

8. **Print your graph** and affix it over the previous axes. Indicate with double arrows on the graph the time periods when the metal was being rubbed.

9. Use the **analysis feature** of the software to read the data from your graph and fill in Table 1-1.

> **Comment:** By rubbing with the same force over the same length, the rate of doing work (the *power*) should be constant. If this is so then the total work done in rubbing the metal on the emery paper should be proportional to the time interval of rubbing. That is, if you rub for twice as long, you will do twice as much work.

Table 1-1

Time interval of rubbing (s)	Final temp. (°C)	Initial temp. (°C)	Temp. change (°C)
20			
40			

Question 1-3: Based on your data, does there appear to be a relationship between the temperature change and the work done by rubbing? Explain.

Question 1-4: What does the shape of your graph imply about the relationship between the work done and the temperature change?

Question 1-5: Why not hold the metal directly instead of using the foam insulation?

Comment: As you rub the metal on emery paper, you are doing mechanical work, but the final result is heat transfer. Thus, it seems appropriate to associate heat transfer as *heat energy transfer.*

It is important to carefully distinguish between the concepts of temperature and heat energy transfer as we have refined them in this activity. They are summarized below.

1. *Heat energy* is energy in transit between two systems in thermal contact due to temperature difference only, with the hotter system losing heat energy as the cooler system gains it. To remind ourselves of this, we will often use the phrase *heat energy transfer.*

2. Two objects are in *thermal equilibrium,* and hence have the *same temperature,* if no energy on average is exchanged between them when they are placed in thermal contact.

If you have additional time, carry out Extension 1-2, in which you will examine temperature changes as a result of conversion of electrical energy to heat energy.

Extension 1-2: Electrical Work and Temperature Change

The hand-operated Genecon generator produces an electric voltage when the crank is turned. The electrical power output of the Genecon (when it is connected to an electrical device like a heater, for example) increases as the crank is turned faster. (Up to a limit!) That is, the Genecon changes mechanical energy into electrical energy. Therefore, rotating the crank at a uniform rate produces electrical energy at a uniform rate (constant *power*).

[If a Genecon generator is not available, you may use a 6-V lantern battery instead. In this case the electric voltage is produced by chemical reactions inside the battery—Chemical energy is changed into electrical energy. As long as the battery is connected to an electrical device like a heater, it produces electrical energy at a uniform rate (constant *power*).]

In addition to the materials above, you will need

- Genecon or other hand-operated electrical generator [or a 6-V lantern battery]

- 47-Ω resistor

- $\frac{1}{2}$-inch-thick slab of foam insulation

1. Explore the effect of the Genecon [or battery] on the resistor. Connect the wires from the Genecon [or battery] to the wires from the resistor. Hold the resistor between your fingers while your partner rotates the crank as fast and steadily as s/he can [or while the battery is connected] for about 30 s.

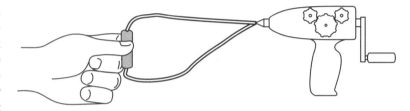

Question E1-6: Describe what you felt. Can electrical energy produce heating effects?

2. The software should be set up as in the previous activity—experiment file called **Mech. and Elect. Work (L2A1-1).**

3. Connect the Genecon to the resistor embedded in the metal. [If you are using a battery, don't connect it until you have been graphing for 10 s.] Insert the temperature sensor in the piece of metal as in Activity 1-1. Keep the metal isolated from the table with the piece of foam insulation.

4. **Begin graphing,** and then after 10 s, begin turning the crank as fast as you can while maintaining a steady rate [or connect the battery]. Crank for 20 s. [Disconnect the battery after 20 s.]

5. After the 20 s, wait until the temperature stops changing. Then wait about 5 s more, *note the time,* and crank for 40 s more. [Connect the battery for 40 s more.] Try to crank at the same steady rate as before.

Stop cranking [disconnect the battery] for the rest of the 180 s.

6. **Print your graph** and affix it over the previous axes. Indicate with double arrows on the graphs the time periods when the Genecon was being cranked [or the battery connected].

7. Use the **analysis feature** in the software to read the data from your graph and fill in Table E1-2.

Question E1-7: Compare your graphs for mechanical and electrical heating. In what ways are they similar and different?

Table E1-2

Time interval of cranking (s)	Final temperature (°C)	Initial temperature (°C)	Temperature change (°C)
20			
40			

Question E1-8:   Does there appear to be a relationship between the temperature change and the electrical energy generated by cranking the Genecon [or by the battery]? Explain.

Question E1-9:   Based on your observations and measurements in this investigation, is it plausible that *heat* is just another form of *energy?* Explain.

Comment: As you crank the generator, mechanical work is causing the generation of electrical energy. Again, as in Activity 1-1, the final result is heat transfer.

In the next investigation you will again use a heat pulser as in Lab 1. Now you can understand the mechanism of the heating effect as the conversion of electrical energy to heat energy.

INVESTIGATION 2: RELATIONSHIP BETWEEN HEAT ENERGY AND TEMPERATURE

If you transfer equal pulses of heat energy to a *perfectly insulated cup* of some liquid, what determines how much temperature change ΔT takes place?

Prediction 2-1:   How does ΔT depend on

A.   The number of pulses of heat energy you transfer (ΔQ)?

B.   The mass (m) of liquid in the cup?

C.   The kind of liquid you have?

In this investigation you will conduct a series of observations in which you examine *quantitatively* the relationship between ΔT and these other variables. To do this you will need to investigate all three factors by changing only one variable at a time. For example, you can use the same mass of room-temperature water for a series of experiments and vary only the amount of heat energy you transfer. Then you can use the same amount of heat energy and vary the mass of the water. Finally, you can use the same mass of liquid and the same amount of heat energy and vary the type of liquid (e.g., use oil instead of water).

To do the series of observations you should have the following equipment:

- computer-based laboratory system with one temperature sensor
- temperature software
- *RealTime Physics Heat and Thermodynamics* experiment configuration files
- heat pulser with 200-W immersion heater
- Styrofoam cup
- stirring rod
- glass beaker (to keep the Styrofoam cup from tipping)
- room-temperature water
- electronic balance
- vat (to prevent spills)

Warnings:

1. **Do not plug in the immersion heater unless it is immersed in water.**

2. **Use enough liquid in each case to make sure the electric coil is just covered in every observation. Be careful not to use large amounts of liquid, because the heating process will take too long! Keep stirring the liquid at all times.**

Activity 2-1: Transferring Different Amounts of Heat Energy to the Same Mass of Water

1. Put a temperature sensor and the heater into the foam cup. (Put the foam cup in the beaker to avoid spillage.) Add 75 g (mL) of room-temperature water to the cup. Be sure that the water covers the coils of the heater.

2. Open the experiment file called **Heating Water (L2A2-1)** to display the axes that follow. This will also set up the heat pulser to transfer 5-s pulses of heat energy.

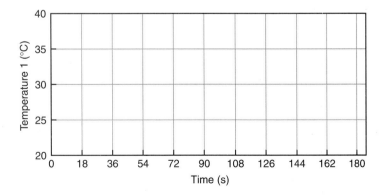

3. **Load the calibration** for the temperature sensor, if this hasn't already been done.

4. **Begin graphing.** *Stir the water vigorously the entire time the computer is graphing the temperature*, now and during the rest of these activities.

5. Measure the initial temperature of the water and record it in Table 2-1. (The temperature can be read more accurately from the digital display than from the graph.)

6. For the first 10 s, don't pulse the heater. Then pulse the heat pulser by pushing the **HEAT** button or key every 10 s for a total of 4 pulses. *Keep stirring.*

7. After the temperature stops changing, record the highest temperature reached as the final temperature in the table.

8. Transfer the data so that the graph will **remain persistently displayed on the screen** for later comparison.

9. Calculate the temperature change and the temperature change per pulse and record these in the table.

Question 2-1: Describe the shape of your graph. What does this say about the relationship between the temperature change and the quantity of heat energy transferred to the water? (Remember that heat pulses were transferred at a constant rate.)

10. Replace the water in the cup with 75 g (mL) of room-temperature water. Record the beginning temperature of the water in Table 2-1.

11. Repeat this activity, transferring twice as many heat pulses (8) at a relatively constant rate to the same amount of water. Use the same setup as before.

    Remember to stir the water continuously while graphing. (Don't pulse the heater during the first 10 s, and then pulse the heater every 10 s for a total of 8 pulses.) When the temperature stops changing, record the final temperature.

12. Calculate the temperature change and change per pulse and record these in the table.

13. **Print your graphs** and affix them over the previous axes.

Question 2-2: How does the change in temperature (ΔT) appear to depend on the amount of heat energy transferred (ΔQ) to a fixed mass of water? State a mathematical relationship in words and as an equation, based on your measurements in this activity.

Table 2-1

Mass of water (g)	Initial temperature (°C)	Final temperature (°C)	Temperature change (°C)	Number of heat pulses transferred	Temperature change per pulse (°C/pulse)
75				4	
75				8	
150					

Question 2-3: Does the temperature *change* produced by one pulse depend on how warm the water is? Why or why not?

Prediction 2-2: Suppose you transfer heat energy to a larger mass of water. How will the temperature change?

A. You just heated 75 g of water with 4 heat pulses. How many pulses do you think it will take to produce the same temperature increase if you heat *twice as much water* (150 g)? _____

B. What will be the temperature increase per pulse if you produce the same temperature increase for *twice as much water* (150 g)? _____

14. Replace the water in the cup with 150 g (mL) of room-temperature water. Record the beginning temperature of the water in Table 2-1.

15. Repeat this activity, transferring enough heat pulses at a relatively constant rate to produce the *same temperature change as produced by 4 pulses for 75 g of water.* Use the same setup as before.

    Remember to stir the water continuously while graphing. (Don't pulse the heater during the first 10 s, and then pulse the heater every 10 s.) When the temperature stops changing, record the final temperature.

16. Calculate the temperature change and change per pulse and record these in Table 2-1.

17. **Print the graph** and affix it below the previous graphs.

Question 2-4: Did the number of pulses required to heat 150 g of water agree with your prediction? Explain.

Question 2-5: Did the rise in temperature per pulse you calculated agree with your prediction? Explain.

Question 2-6: Based on your graphs and data, does the following mathematical relationship make sense?

$$\Delta Q = cm\, \Delta T$$

ΔQ is the heat energy transferred to the water, m is the mass of the water, ΔT is the change in temperature, and c is a constant characteristic of the liquid. Explain how well your results are described by this relationship.

In the previous two activities you have examined the relationship between the amount of heat energy transferred to a system and the system's change in temperature. You have seen that the change in temperature is proportional to the amount of heat energy transferred and inversely proportional to the mass of the system. To be more quantitative (e.g., to be able to predict numerical temperature changes), it is necessary to specify what amount of heat energy transfer will produce a one degree change in temperature in unit mass of a material. This quantity is known as the *specific heat* of the material. It is the value c in the equation in Question 2-6.

$$\text{specific heat} = c = \frac{\Delta Q}{m \, \Delta T}$$

The standard units for heat energy (J), mass (kg), and temperature (°C), give us the unit for specific heat, J/kg-°C.

In the next activity you will calculate the specific heat of water from your data.

Activity 2-2: Specific Heat of Water

1. Enter the number of pulses and temperature change data from Activity 2-1 in Table 2-2. Carry out the calculations in steps 2–3 to fill in the rest of the table.

2. Calculate the total heat energy transferred by the heater using the power rating of the heater in watts (W) and the total time of heat pulses transferred to the water in each run of the experiment. (Recall that 1 W = 1 J/s.)

3. Calculate the specific heat for each run.

4. Calculate the average value of the specific heat of water from the three values in your table:

$$c_{\text{water}} =$$

Question 2-7: How closely did the three values of the specific heat agree with each other? How did the average value agree with the accepted value, $c_{\text{water}} = 4190$ J/kg-°C? What are the possible sources of experimental error that might explain any disagreement?

Table 2-2

Mass of water (kg)	Number of heat pulses	Total time of heat pulses (s)	Total heat energy transferred by heater (J)	Change in temperature (°C)	Specific heat (J/kg-°C)
0.075	4				
0.075	8				
0.150					

Question 2-8: If the heater were plugged directly into a wall outlet so that it is transferring heat to the water continuously, how many seconds would it take to raise the temperature of 300 g of water by 25°C? Show your calculation and explain your reasoning. (Use $c_{water} = 4190$ J/kg-°C.)

If you have enough time, carry out Extension 2-3 to find the specific heat of another liquid, vegetable oil.

Extension 2-3: Specific Heat of Vegetable Oil

You can use the same method as in Activities 2-1 and 2-2 to determine the specific heat of a different liquid—vegetable oil. In addition to the materials used before, you will need

- room-temperature vegetable oil
- clean, dry Styrofoam cup

Warning: Do not heat the oil over 70°C at any time during the experiment!!!!

1. Follow the same procedure as in Activities 2-1 and 2-2 to find the specific heat of vegetable oil. Use just one run with a sample of about 75 g. (*Note that vegetable oil has a different density than water.*)

2. Describe the steps below *in words* and record all data, calculations, and the measured specific heat below.

Description of the procedure:

Data:

Calculations:

$$c_{oil} = \underline{\hspace{1cm}}$$

Question E2-9: Which substance changes temperature the most for a given amount of heat energy transfer, the water or the oil? Which has the larger specific heat?

Question E2-10: Back in the good old days when a kid was given a hot baked potato to carry to school on a cold winter day, the potato kept her warm and served as lunch! Why is a hot potato a better kid warmer than a bag of hot popcorn?

INVESTIGATION 3: THE MECHANICAL EQUIVALENT OF HEAT

Before the mid-nineteenth century, *heat* was regarded as a substance rather than as a form of energy exchange between two substances. Heat was measured in its own special units called *calories*. By definition,

1 calorie = the quantity of heat that raises the temperature of
1 gram of water by 1 degree Celsius

(Note that the food calorie (also called a kilogram calorie or kilocalorie) is 1000 times larger than this.)

According to the definition of the calorie, the specific heat of water is

$$c_{water} = 1.0 \text{ cal/g-°C}.$$

In the mid-nineteenth century, James Joule carried out a series of experiments converting mechanical and electrical energy to heat, demonstrating that heat is a form of energy and not a substance, as you have seen in Investigation 1.

The immersion heater that you used in Investigation 2 works in the same way as the resistor connected to the Genecon generator (or battery) used in Investigation 1. The electrical energy transferred to the heater (supplied through the electrical outlet) is produced by a generator located at a power plant. The equivalent of your hand turning the crank of the Genecon is the transformation of either chemical potential energy (stored in fossil fuels), nuclear potential energy (stored in nuclear fuels), or gravitational potential energy (stored in water above a dam) into mechanical energy of the generator.

In the next activity you will find the quantitative *mechanical equivalent of heat*. Thus you will determine the number of joules that are equivalent to a calorie of heat energy.

Activity 3-1: The Energy Equivalent of the Calorie

1. Choose the run in Table 2-2 that gave the specific heat value closest to 4190 J/kg-°C and enter the data in Table 3-1.

2. Calculate the total heat energy transferred to the water *in calories* using the relationship $\Delta Q = cm \, \Delta T$, using the specific heat of water in cal/g-°C, and enter in the fifth column of the table.

3. Calculate the mechanical equivalent of heat in J/cal.

HOMEWORK FOR LAB 2:
ENERGY TRANSFER AND TEMPERATURE CHANGE

1. How is it possible that rubbing a piece of metal on emery paper can result in an increase in the temperature of the metal? How is it possible that running an electric current through a piece of metal can result in an increase in temperature? In each case, what is being transferred to the metal?

2. Explain on the basis of your observations in this laboratory, how you know that heat is a form of energy. Explain what evidence you found that heat is not a *substance* that flows from a hotter body to a colder one. Refer to specific observations. What is special about the form of energy known as heat?

3. If you transfer heat energy to a perfectly insulated cup of some liquid (no heat energy can be transferred in or out through the walls), what determines how much the temperature changes? Does it depend on how much heat energy you transfer, how much liquid there is in the cup, what the liquid is, or what the initial temperature of the liquid is? Which of these factors do you think make a difference in how much the temperature rises? (Give evidence from your observations.)

4. Suppose you transfer a certain amount of heat energy to a known amount of a liquid in a perfectly insulated cup and the temperature changes. Then, you decide to alter the experiment in several different ways. For each of the alterations listed below, state whether the total change in temperature will be larger, smaller, or the same as that measured in the original experiment (before the alteration). Briefly explain each answer.

   a. Transfer more heat energy.

b. Heat for a longer time but transfer the same total heat energy.

c. Start with more liquid in the cup.

d. Increase the starting temperature of the liquid.

e. Use an equal mass of liquid that has a larger specific heat.

f. Use the same volume of a denser liquid that has the same specific heat.

5. You have two perfectly insulated cups. One contains water and the other contains an equal volume of another liquid that has half the density of water and two times the specific heat. You heat the water from 10 to 20°C and the other liquid from 80 to 90°C. Compare the amount of heat energy transferred to raise the temperature of the other liquid to the amount transferred to raise the temperature of the water. Explain how you got your answer.

6. A perfectly insulated cup is filled with water initially at 20°C. Heat energy is transferred to the cup by an immersion heater at a steady rate. Sketch on the axes that follow the temperature as a function of time. Explain the shape of your graph and write down a mathematical relationship for the temperature (T) as a function of the quantity of heat energy (ΔQ) that has been transferred to the water.

318

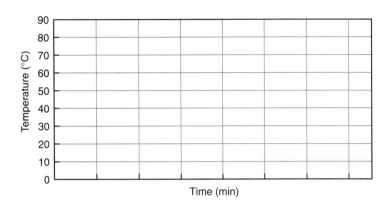

7. How many calories of heat energy would it take to heat 200 mL of water from 20 to 80°C? Show your calculations.

8. Compare the heat energy needed in Question 7 to that needed to heat 100 mL of water from 60 to 90°C? Explain your answer.

9. How long would it take to heat the water in Question 7 using a heater with a power rating of 50 W? Show your calculations.

10. A piece of aluminum (specific heat 910 J/kg°C) of mass 200 g at 80.0°C is dropped into a Styrofoam cup filled with 100 mL of water at 20°C. What are the final temperatures of the water and the aluminum? Show your calculations.

Pre-Lab Preparation Sheet for Lab 3:
Heat Energy Transfer

(Due at the beginning of Lab 3)

Directions:
Read over Lab 3 and then answer the following questions about the procedures.

1. What is your Prediction 1-2?

2. How will you test your Prediction 1-2 in Activity 1-1?

3. In Activity 2-3, why are both cups covered?

4. What is your Prediction 3-1? Which can will cool at a faster rate, or will they both cool at the same rate?

5. What is your Prediction 3-2? Which can's temperature will rise faster, or will they both rise at the same rate?

LAB 3:
HEAT ENERGY TRANSFER

R-31 insulation in the attic, R-19 in the walls, insulated glass windows, glass fireplace doors, that should do you this winter . . .

—Local Energy Consultant

OBJECTIVES

- To examine the dependence of heat energy transfer between two objects on the difference in temperature between the objects.

- To explore what other factors influence the rate of heat energy transfer by conduction and convection.

- To examine what types of materials are good conductors of heat energy.

- To explore the transfer of heat energy by emission and absorption of radiation.

OVERVIEW

In Labs 1 and 2 you studied the basic thermodynamic concepts of temperature and heat energy transfer. In this lab you will look more quantitatively at the factors that affect the rate at which heat energy is transferred from one object to another.

   You know that it is more expensive to heat a house in winter than in the spring. In both cases, you try to keep the interior temperature the same, but the outside temperature is often considerably lower in the winter. What effect does this have on the rate of heat energy transfer from inside to outside? In areas where it gets particularly cold in the winter, it is worthwhile to spend considerable amounts of money and effort insulating your home.

What are the most effective ways of slowing the inevitable transfer of heat energy from inside to outside your home in the winter, or outside to inside in the summer? A hot cup of coffee on the table will eventually cool down to room temperature. How can you keep it hot longer? You might try a cover, or a Styrofoam cup, or even a Thermos bottle to slow down the inevitable. Which of these is more effective, and how does each work to slow cooling of the coffee?

In the first investigation you will explore how the rate of heat energy transfer depends on the difference in temperature. This should shed some light on your winter heating bill. In the second and third investigations you will look at the ways that heat energy is transferred—*conduction, convection, and radiation*—and the most effective ways of reducing each of them.

INVESTIGATION 1: HEAT ENERGY TRANSFER AND TEMPERATURE DIFFERENCE

Without perfect insulation, a house will transfer heat energy to its surroundings when it is colder outside than inside. How much energy is needed to replace this heat energy? In this investigation, you are going to explore this question by keeping a cup of water at a constant temperature by replacing the heat energy that is transferred to the surroundings. You have seen in Lab 1 how to do this by transferring heat energy to the water using a heat pulser. In this investigation you will examine how the rate of heat energy transfer from hot water to its cooler surroundings depends on the difference between the temperature of the water and room temperature. You can accomplish this by measuring the number of heat pulses needed to keep water of different temperatures warm for a fixed length of time.

Prediction 1-1 (review): If you transfer heat energy into the water at the same rate as heat energy is being transferred from the water to its cooler surroundings, what will happen to the temperature of the water? What would happen if you transferred heat energy into the water at a faster rate than it is being transferred out? At a slower rate than it is being transferred out?

Prediction 1-2: How does the rate at which you must transfer heat energy to keep the water temperature constant depend on the temperature of the water? Suppose that room temperature is 20°C and you have two cups of water, one at 30°C and one at 60°C. Will you need to transfer heat energy at a different rate to keep these cups at 30 and 60°C for 5 min? If so, which one will require a higher rate of heat energy transfer? How much higher? Twice as large, three times as large, or some other ratio? How did you determine your answer?

To test your predictions you will need

- computer-based laboratory system with one temperature sensor

- temperature software

- *RealTime Physics Heat and Thermodynamics* experiment configuration files

- heat pulser (relay box and immersion heater)

- hot water (about 70°C)

- container marked in mL
- stirring rod
- *uninsulated* glass, plastic, metal, or paper cup
- vat (to prevent spills)

Activity 1-1: Keeping Hot Water Hot

> **Warning!!** Do not plug the immersion heater into the pulser or press the heat button or key unless the heater is in the water. The heater will burn out if it is turned on even for a short time while not immersed in water.

1. Plug the temperature sensor and the heat pulser into the interface.

2. Open the experiment file called **Keeping Hot (L3A1-1)** to set up the axes that follow. This will also set up the heat pulser to transfer 2-s pulses of heat each time the **HEAT** button or key is pressed.

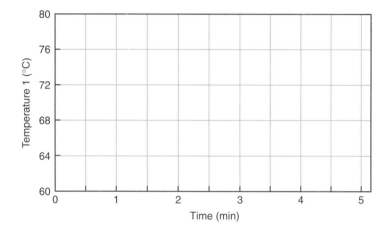

3. **Load the calibration file** for the temperature sensor.

4. Record room (air) temperature in Table 1-1.

5. Pour about 100 mL of hot water about 45°C warmer than room temperature (about 65°C) into the *uninsulated,* uncovered cup. Put the sensor and heater into the water and plug the heater into the relay box just before starting. (Use the heat pulser to warm the water before you start, if necessary.)

Table 1-1

Temperature of water (°C)	Room (air) temperature (°C)	Temperature difference (°C)	Number of heat pulses transferred in 5 min

6. **Begin graphing,** and *while stirring continuously* try to keep the water temperature constant (within a degree) for the full 5 min. Press the **HEAT** button or key when you need to transfer more heat energy to the water. The computer will make marks on the graph each time you pulse the heater and will keep track of the number of pulses you use. *Keep stirring!*

7. When you are done, record in Table 1-1 the number of heat pulses that were transferred.

Question 1-1: About how many heat pulses would you need to keep the same amount of water at the same temperature for ten min? Explain your answer.

Prediction 1-3: How many pulses do you predict it will take to keep the same amount of *warm* water (30°C warmer than room temperature) at a constant temperature for 5 minutes compared to the number you needed for hot water (45°C warmer than room temperature)? Explain your answer.

Test your prediction.

8. Repeat steps 3–6 above, this time for water that is about 30°C above room temperature. You will need to **change the temperature axis** on your graph. Be sure to stir and keep the temperature constant for the full 5 min.

9. Record your data in Table 1-1.

Question 1-2: How did your results compare with your prediction?

Prediction 1-4: How many pulses do you think it will take to keep the same amount of *lukewarm* water (15°C warmer than room temperature) at a constant temperature for 5 minutes compared to the number you needed for hot water (45°C warmer than room temperature)? Explain your answer.

Test your prediction.

10. Repeat steps 3–6 above, this time for water that is about 15°C above room temperature. Be sure to stir and keep the temperature constant for the full 5 min.

11. Record your data in Table 1-1.

Prediction 1-5: How many pulses do you think it will take to keep the same amount of *room-temperature* water at a constant temperature for 5 minutes compared to the number you needed for hot water (45°C warmer than room temperature)? Explain your answer.

Test your prediction.

12. Repeat steps 3–6 above, this time for water that is at room temperature. Be sure to stir and keep the temperature constant for the full 5 min.

13. Record your data in Table 1-1.

Question 1-3: Describe the relationship between rate of heat energy transfer and difference in temperature based on your data in Table 1-1.

Clearly the number of pulses per second needed to keep the water warm (the rate of heat energy transfer) is related to the difference in temperature between the water and room temperature. Recall that this is also the rate at which heat energy is transferred out of the water into its surroundings. If you have additional time, carry out Extension 1-2 in which you can try to find a mathematical relationship between rate of heat transfer and temperature difference.

Extension 1-2: The Mathematical Relationship Between Heat Energy Transfer and Temperature Difference

Prediction E1-6: Based on your results in Activity 1-1, what do you think is the mathematical relationship between the rate at which heat energy is transferred from the water in the cup into the surroundings and the difference between the temperature of the water and room temperature? Explain the basis for your prediction.

Test your prediction.

1. Open the experiment file called **Mathematical Relation (L3E1-2)** to display a table and active graph for number of heat pulses vs. temperature difference.

2. **Enter the data** from Table 1-1 to plot a graph.

3. Determine the mathematical relationship between the rate of heat energy transfer and the temperature difference using the **fit routine** in your software.

4. **Print the graph** and affix it in the space below.

Question E1-4: What is the mathematical relationship between the rate of heat energy transfer from a sample of hot water and the difference between the temperature of the water and room temperature? Explain how you arrived at your answer.

Question E1-5: The temperature inside a house is kept at a constant 20°C. Use your answer to Question E1-4 to compare the cost of heating the house on a spring day when it is 10°C outside to that of heating the house on a winter day when it is −10°C outside. (Assume that the cost per unit of heat energy transferred by the furnace is always the same.)

INVESTIGATION 2: CONTROLLING THE TRANSFER OF HEAT ENERGY

In the previous investigation you observed that the rate at which heat energy is transferred from a warmer object to a cooler one is roughly proportional to the *difference* in temperature. Consider a cup of hot coffee on the table in the lab. You might put it in a Styrofoam cup with or without a lid, a ceramic cup, a metal cup, etc. In each case, the initial temperature difference is the same, and in each case we know that the coffee will eventually cool down to room temperature. What else does the rate at which the temperature of an object changes depend on?

The transfer of heat energy from the hot coffee to its surroundings takes place through several different mechanisms. There is *conduction* through the cup into the table and surrounding air. There is *convection* from the air flow that results when the air is heated by the cup and coffee. Convection can also include some evaporation from the surface of the coffee. Finally, there is *radiation*, the electromagnetic waves (mostly infrared) emitted from the hot surfaces of the cup. In this investigation you will do several experiments to see whether conduction or convection is a more important mechanism of heat energy transfer. In the next investigation, you will look at radiation.

Prediction 2-1: Which cup do you think will cool faster, a covered cup or an uncovered cup, both containing water at the same initial temperature? Do you think that covering a cup will make much difference in the rate of cooling?

You will need the following to test your prediction:

- computer-based laboratory system with two temperature sensors
- temperature software
- *RealTime Physics Heat and Thermodynamics* experiment configuration files
- hot water (about 80°C)
- container marked in mL
- immersion heater (optional, for heating water)
- 2 Styrofoam or other insulated cups with 75-mL marks
- cover with small hole for temperature sensor
- vat (to prevent spills)

Activity 2-1: To Cover or Not to Cover?

1. Plug the two temperature sensors into the interface.

2. Open the experiment file called **Cooling Down (L3A2-1)** to set up the axes that follow.

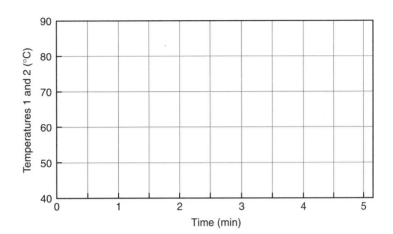

3. **Load the calibration file** for the two temperature sensors.

4. Record room (air) temperature: _____

5. Place the two foam cups on the table and be ready to cover one. Push the temperature sensor through the hole in the cover so that it fits tightly. Be sure that the cover fits tightly and that there are no other holes in it.

6. Pour the same amount of water (about 75 mL) at 80°C into each cup. This can most easily be done quickly by making a 75-mL mark on both cups beforehand and then filling to this mark.

7. Now, put the cover (with sensor) on one cup. Put a sensor in the other cup. *Both cups should be at the same temperature—80°C—when you begin graphing.* If they are not, use the immersion heater to heat the water.

8. **Begin graphing** and continue for the full 5 min.

9. **Print your graph** and affix it over the axes.

10. Use the **analysis feature** in the software to record the initial and final temperatures of each cup in Table 2-1.

Question 2-1: From which cup was the most heat energy transferred during the 5 min? If you had to keep these cups at the same constant temperature using a heater, to which cup would you need to transfer heat energy at a faster rate?

Question 2-2: Did the cover slow down the rate of heat energy transfer very effectively? Explain your answer, using your experimental results.

Table 2-1

Experimental conditions	Initial temperature (°C)	Final temperature (°C)	Temperature change in 5 min (°C)
Covered foam cup			
Uncovered foam cup			
Covered foam cup			
Covered metal cup			
Covered foam cup			
Covered double foam cup			
Covered shiny metal can			
Covered black painted can			

Prediction 2-2: You found that a cover slowed down the rate of heat energy transfer. Which cup do you think will cool faster, a covered metal cup or a covered foam cup? Do you think that one will be very different in its rate of cooling?

To test your prediction you will need the following materials in addition to those from the previous activity:

- metal cup about the same size as the Styrofoam cup
- tightly fitting cover with hole for temperature sensor

Activity 2-2: Metal or Foam Cup

Before you carry out this activity, fill up the metal cup and the foam cup with hot water, wait awhile, and then touch the side of each.

Question 2-3: Do the sides of the two containers seem to have the same temperature? If they feel different, which one feels hotter?

Compare the rate of temperature decrease (and heat energy transfer) through a *covered* metal cup and through a *covered* foam cup.
Use 80°C water and the method of the previous activity. *Be sure that both starting temperatures are the same, and that the covers fit tightly.* Again graph for 5 min.
Record your data in Table 2-1.

Question 2-4: Which cup had the smaller change in temperature during the 5 min? If you had to keep these cups at the same constant temperature using a heater, to which cup would you need to transfer heat energy at a faster rate?

Prediction 2-3: Which cup do you think will cool faster, a *covered* double-foam cup or a *covered* single-foam cup? Do you think that one will be very different from the other in its rate of cooling?

Test your prediction.

Activity 2-3: Covered Double-Foam and Covered Single-Foam Cups

Compare the rate of temperature decrease and the rate of heat energy transfer through a *covered* double-foam cup and through a *covered* single-foam cup. Use 80°C water and the method of the previous activities. *Be sure that both starting temperatures are the same and that the covers fit tightly.*

Record your data in Table 2-1.

Question 2-5: Which cup had the smaller change in temperature?

Use your data in Table 2-1 to answer the following questions.

Question 2-6: Which cup seemed most effective in decreasing the rate of heat energy transfer to the surroundings? Least effective? If you wanted to keep a cup of coffee hot, what is the most important thing to do?

1. cover the container

2. substitute foam for metal

3. double the thickness of the foam

Describe the data that support your conclusions.

Question 2-7: Why was it important to cover all of the cups after the first activity?

Question 2-8: Which mode of heat energy transfer—*conduction* or *convection*—is most affected by each of the methods explored in this investigation?

1. covering the container

2. substituting foam for metal

3. doubling the thickness of the foam

If you have extra time, see if you can solve the puzzler in Extension 2-4.

Extension 2-4: Puzzler: Which Feels Cooler?

According to your observations in this and the previous two labs, objects left in thermal contact for a long time will tend to reach thermal equilibrium, i.e., end up at the same temperature. Therefore, if you have a piece of foam and a piece of metal sitting around the lab for awhile, they should be the same temperature. For the next activity you will need the following materials in addition to your computer-based temperature sensor:

- slab of metal with hole for temperature sensor
- slab of Styrofoam with hole for temperature sensor

Prediction E2-4: Are objects lying around a room really at the same temperature? Feel the metal and Styrofoam with your fingers. Predict which one actually has the highest temperature and the lowest temperature.

Measure the temperature of the metal and the Styrofoam with the temperature sensor and record your measurements in table E2-4. (Be sure that the sensor is in contact long enough so that thermal equilibrium has been reached.)

Question E2-9: Did your observations agree with your prediction? Is your sense of touch an accurate predictor of relative temperatures?

Question E2-10: According to other observations you have made, should the temperatures of the two different materials sitting around in the same room be the same or different?

Question E2-11: On the basis of observations you have made of heat energy transfer by conduction in Activity 2-2, you should be able to explain the reason why some objects *feel* colder than others. (**Hints:** Is the temperature of your hand the same as room temperature? What happens when you touch an object that is at room temperature? Which objects are thermal conductors? Thermal insulators?)

Table E2-4

Material	Temperature (°C)
Styrofoam	
Metal	

INVESTIGATION 3: HEAT ENERGY TRANSFER BY RADIATION

In the previous investigation, you examined the transfer of heat energy by conduction and convection. You looked at several ways that this transfer can be controlled by changing the construction of the container and the materials from which it is made. In this investigation you will examine how the properties of the surface of the container affect radiation, the third important means of heat energy transfer.

First a prediction.

Prediction 3-1: Two covered cans are filled with hot water. One is an unpainted (shiny) metal can, while the other is a dull black painted metal can. Which water do you think will cool at a faster rate? Do you think that one will be very different from the other in its rate of cooling? Why?

To test your prediction you will need

- computer-based laboratory system with two temperature sensors
- temperature software
- *RealTime Physics Heat and Thermodynamics* experiment configuration files
- hot water (about 80°C)
- container marked in mL
- immersion heater (optional, for heating water)
- two identical cans, one unpainted (shiny) and one painted flat black on the outside
- tightly fitting covers with holes for temperature sensor
- 2 pieces of foam insulation about 10 cm square and 2 cm thick
- vat (to prevent spills)

Activity 3-1: Shiny Metal Can and Dull Black Painted Can

Compare the rate of temperature decrease (and the rate of heat energy transfer) through a *covered* shiny metal can and through a *covered* dull black painted can. Place the cans on top of the foam slabs. (The covers control heat energy transfer by convection and the foam pads control heat energy transfer by conduction into the table.)

Use 80°C water and the method of the Activities 2-1 to 2-3. *Be sure that both starting temperatures are the same and that the covers fit tightly.*

Record your data in Table 3-1 and Table 2-1.

Question 3-1: Which cup had the smaller change in temperature in the 5 min? If you had to keep these cups at the same constant temperature using a heater, to which cup would you need to transfer heat energy at a faster rate?

Table 3-1

Experimental conditions	Initial temperature (°C)	Final temperature (°C)	Temperature change in 5 min (°C)
Covered shiny metal can			
Covered black painted can			

You observed that a covered shiny can cools down somewhat more slowly than an identical can painted on the outside with flat black paint. What is going on here? It can't have to do with convection or conduction, since these have been reduced by the cover and the foam pad, and, in any case, should be the same for both cans. In the next activity, you will further examine the transfer of energy by radiation and how it is affected by the surfaces of objects.

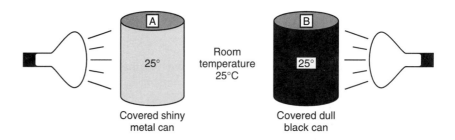

Prediction 3-2: Two cups of water absorb radiation from identical heat lamps. Both cups are covered and are identical except that one is shiny and the other is painted with flat black paint on the outside. Both cans contain 100 g of water at room temperature. Will the temperature of the water rise faster in the shiny or black can? Explain.

To test your prediction you will need the following materials in addition to those from the previous activity:

- 250-W heat lamp and socket
- ring stand and clamp
- meter stick
- piece of white paper to shield the temperature sensor

Activity 3-2: Heating With Radiation

1. Plug one temperature sensor into the interface.

2. Open the experiment file called **Radiation (L3A3-2)** to set up the axes that follow.

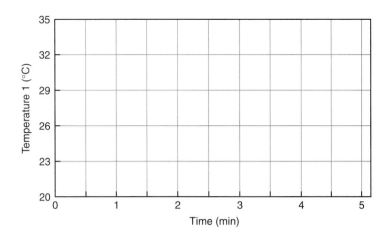

3. **Load the calibration file** for the temperature sensor.

4. Set up the heat lamp and black can as shown below, with the can on top of the foam pad. The face of the lamp and the can should be directly across from each other at the same height, and about 35 cm apart. The paper shield should be taped so that it prevents radiation from the lamp from reaching the temperature sensor or wire directly.

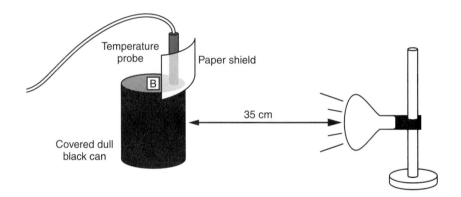

5. Pour 75 mL of room-temperature water into the can, insert the temperature probe in the cover, and place the cover tightly on the can.

6. Record the initial temperature of the water in Table 3-2.

7. **Begin graphing** and turn on the heat lamp.

8. At the end of 5 min, when the graphing is over, record the final temperature of the water in the table.

9. Move the data so that the graph is **persistently displayed on the screen.**

Table 3-2

Experimental conditions	Initial temperature (°C)	Final temperature (°C)	Temperature change in 5 min (°C)
Black covered metal can			
Shiny covered metal can			

10. Repeat the procedures in steps 5–8 with the shiny metal can at exactly the same distance from the heat lamp. Record the initial and final temperatures in the table.

11. Calculate the temperature changes for 5 min for the two cans from your data.

12. **Print your graphs** and affix them over the previous axes.

Question 3-2: How did your observations agree with your prediction?

Question 3-3: Based on your observations, to which surface does the radiation from the heat lamp seem to be transferred at a faster rate—the black one or the shiny one?

Question 3-4: Based on your observations in Activity 3-1 and your answer to Question 3-1, what is the relationship between effectiveness at emitting radiation and effectiveness at absorbing radiation? If a surface is effective at absorbing radiation, will it be effective at emitting the same radiation?

HOMEWORK FOR LAB 3:
HEAT ENERGY TRANSFER

1. Based on your observations, which of the following seems to most accurately predict how fast heat energy is transferred from one object to another? Explain the experimental basis for your answer.

    a. the temperature of the hotter object

    b. the temperature of the colder object

    c. the difference in temperature between the two objects

2. The air temperature outside on a winter day is 10°C. Two identical outdoor swimming pools are heated by electric heaters. Swimming pool A is maintained at 20°C, while pool B is maintained at 30°C. Compare the rate at which heat energy is transferred to pool A with the rate at which heat energy is transferred to pool B. Be quantitative (give a ratio for the power ratings of the two heaters) and explain your answer in terms of your observations in this laboratory.

3. Suppose that the outside temperature drops to 5°C. Now compare the new rate at which heat energy must be transferred to pool A to maintain it at 20°C to the rate in Question 2. Again be quantitative and explain your answer.

4. Room temperature is 20°C. For what temperature of an object will there be no net heat energy transfer from the object to the room or from the room to the object? Explain your answer.

5. If objects sitting in a room for a long time are all at the same temperature, why do metal objects feel cooler to the touch than cloth or plastic ones?

6. Which of the following changes is most effective in reducing heat energy transfer from a container of hot water, and which is least effective?

   a. Replacing a metal can with a black surface with one with a shiny surface

   b. Replacing a metal can with a foam cup

   c. Replacing a single foam cup with two nested foam cups

   d. Putting a thin plastic cover on the open cup

   e. Replacing a thin plastic cover with a foam cover

   Use your results to explain and support your answers.

7. Summarize the differences between the three heat energy transfer processes examined in this lab: conduction, convection, and radiation.

8. Which of the changes in Question 6 (a–e), will *most* affect each of the three heat energy transfer processes listed below? Briefly explain each answer. (Note that each change may affect more than one of the processes.)

   Conduction

   Convection

   Radiation

9. Based on your answers to Questions 6, 7, and 8, which heat energy transfer process (conduction, convection, or radiation) is most important in the cooling of an uncovered, uninsulated cup of hot water (80°C or more)? Explain your answer.

10. Suppose that you had the following materials: metal can, Styrofoam cup, sheet of plastic wrap, plastic lid, sheet of aluminum foil, flat black paint. You have 200 mL of water at 80°C. Describe a container made from these materials that will keep the water warm as long as possible.

11. Assuming that a roofing material has the same emission and absorption properties for visible and infrared radiation, would you choose a dark- or light-colored roof if you lived in a climate that is (a) warm and very sunny all year round, and (b) cold and cloudy for much of the year? Explain based on your observations in this lab.

12. A Thermos bottle has the following features to control transfer of heat energy in and out: (a) a tight cover, (b) a shiny metallic outside coating, and (c) an air layer between the inside and outside surfaces. What function does each of these features serve in controlling heat energy transfer?

Name_____   Date_____

PRE-LAB PREPARATION SHEET FOR LAB 4:
THE FIRST LAW OF THERMODYNAMICS

(Due at the beginning of Lab 4)

Directions:
Read over Lab 4 and then answer the following questions about the procedures.

1. Sketch your graph for Predictions 1-1 and 1-2 on the axes below using solid and dashed lines.

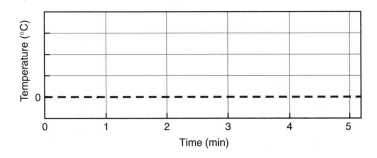

2. Why is a Styrofoam (insulated) cup used in Activity 1-1?

3. Sketch your Prediction 2-1 on the axes below.

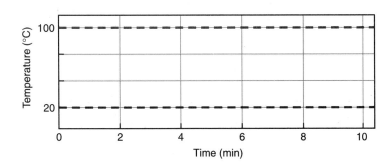

4. What is the purpose of the two different size syringes in Activity 3-2?

5. What is your Prediction 3-1?

LAB 4:
THE FIRST LAW OF THERMODYNAMICS

It must be admitted, I think, that the laws of thermodynamics have a different feel than most of the other laws of the physicist. There is something more palpably verbal about them—They smell more of their human origin.

—P.W. Bridgman

OBJECTIVES

- To understand phase changes as processes in which energy is transferred to a system without the temperature increasing, and to understand the concept of latent heat.

- To determine the latent heats of fusion and vaporization of water.

- To understand the meaning of *internal energy* of a system.

- To understand the energy balance of a system between heat energy transfer, work done, and changes in internal energy, as described by the first law of thermodynamics.

OVERVIEW

In Labs 1–3 you studied the basic thermodynamic concepts of temperature and heat energy transfer. You learned how heat energy can be transferred from one system to another. In this lab we will extend the idea of heat energy transfer to changes in the motions and internal conditions of atoms that scientists believe make up matter. We will also examine mechanical work as an alternative way of adding or removing energy from a system.

In the first investigation you will consider how the transfer of heat energy can change a system internally without changing its temperature. The process you will examine is the change of a substance from solid to liquid, called a *change of phase*. As part of this activity, you will measure the latent heat of fusion of water—the amount of heat energy transfer required to transform one unit of mass of ice to liquid water.

Next you will consider a second example of a process in which heat energy transfer changes a system internally without changing its temperature. This is a phase change from liquid to gas. You will measure the latent heat of vaporization of water—the amount of heat energy transfer required to transform one unit of mass of water from liquid to steam.

Finally, you will explore the work associated with a gas that expands or is compressed in a cylinder with a piston. This will lead to the idea that the energy stored internally in a system can be increased by either transferring heat energy to it *or* by doing work on it. The law that keeps track of the *internal energy* of a system, the *work done by the system,* and *heat energy transferred to the system* is the first law of thermodynamics. An understanding of this extended conservation of energy law is important in practical endeavors, such as the design of heat engines, and fundamental research into the nature of the universe.

INVESTIGATION 1: HEAT ENERGY TRANSFER WITHOUT TEMPERATURE CHANGE

As part of our quest to understand more about heat energy transfer and the internal energy of a substance, let's look at the question of whether net heat energy transfer can take place without a temperature change occurring. Just in case you haven't memorized this yet, let's return to the principle of heat energy transfer we developed in Labs 1-3:

Heat is energy that is transferred between two systems in thermal contact with each other due to a temperature difference between them.

Remember that although the term "heat" is used, even by scientists, it is a shorthand term for an energy transfer process rather than a substance transfer. We will continue to use consistently the term *heat energy transfer* to remind you that the transfer of heat from one system to another does not necessarily involve the exchange of matter.

Let's return to the question at hand. Is it possible for a net amount of heat energy to be transferred to a system without a change in temperature?

Prediction 1-1 (review):   On the axes below, sketch a graph of temperature vs. time for a cup of cool (liquid) water (initially at 0°C) heated in the following way:

- First 30 s, no heat energy transferred.

- Next 3 min, heat energy transferred at a constant rate.

- Next 1 min and 30 s, no heat energy transferred.

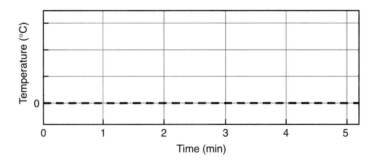

Explain the shape of your graph based on what you know about the relationship between heat energy transfer and temperature.

Prediction 1-2: How would the temperature history above be different if the cup contained a mixture of ice and water at 0°C? Describe any differences, and sketch this temperature history on the same axes using a dashed line.

To test your prediction, you will need

- computer-based laboratory system with one temperature sensor
- temperature software
- *RealTime Physics Heat and Thermodynamics* experiment configuration files
- heat pulser (relay box and 200-W immersion heater)
- crushed ice and cold water
- container marked in mL
- stirring rod
- Styrofoam cup, 300 mL, (for crushed ice)
- vat (to prevent spills)
- electronic balance

Activity 1-1: Heating Ice and Water: A Temperature History

> **Warning!!** Do not plug the heater into the pulser or press the HEAT button or key unless the heater is in the water. The heater will burn out if it is turned on even for a short time while not immersed in water.

1. Plug the temperature sensor and the heat pulser into the interface.

2. Open the experiment file called **Ice to Water (L4A1-1)** to set up the axes that follow. This will also set up the heat pulser to transfer 10-s pulses of heat energy each time the **HEAT** button or key is pressed.

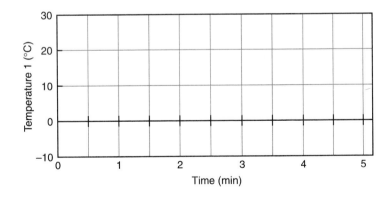

3. **Load the calibration file** for the temperature sensor.

4. Put the temperature sensor and heater in the foam cup. Let the heater sit on the bottom of the cup. When you are ready to begin, add 75 g (mL) of water *that has been cooled to near 0°C.*

Get some crushed ice, blot it on a paper towel to remove any water drops, and *immediately* add to the cup an amount of ice about equal in volume to the amount of water already present.

5. *Stir vigorously with the stirring rod, and continue to do so for the rest of this experiment.* Measure the beginning temperature of the ice/water mixture, and record it in Table 1-1.

6. **Begin graphing** and **pulsing the heat pulser** one pulse after another, at a constant rate. *(As soon as one pulse stops, immediately press the **HEAT** button or key to begin the next pulse.) Keep stirring vigorously!* Watch the mixture and record in Table 1-1 the time when all of the ice is melted, and the number of heat pulses added up until this time.

   Continue to stir and to pulse the heater.

7. Continue pulsing the heater and stirring until the temperature is about 10°C above the beginning temperature. *Keep stirring* even after you stop pulsing the heater, until the temperature stops rising.

8. Measure the final amount of water in the cup and calculate the mass of ice that was added. Record these in Table 1-1.

9. **Print your graph** and affix it over the previous axes. Indicate with an arrow the time when the ice had completely melted.

Question 1-1:   Does your graph agree with the prediction you made for an ice and water mixture? If not, describe in words the ways in which the temperature history of the ice and water mixture was different than the one for water alone that you sketched in Prediction 1-1.

Question 1-2:   During the time that the temperature remained constant, what do you think happened to the heat energy you were transferring to the water and ice mixture if it wasn't raising the temperature?

> **Comment:**   When heat energy is transferred to a mixture of ice and water in thermal equilibrium at the melting point temperature, all of the heat energy transferred is absorbed by the ice and used to break down the rigid bonds holding the molecules together in the ice crystal structure. This represents a very significant change in the internal structure, and internal energy of the system. Yet, while this is going on, *the temperature of the mixture does not increase.*

The heat energy transfer required to melt one kilogram of ice when its temperature is already at its melting point is called the *latent heat of fusion.* (This amount

Table 1-1

Initial mass of water (kg)	Initial temperature (°C)	Time when all ice had melted (s)	Number of heat pulses to melt ice	Total time of heat pulses (s)	Final mass of water plus ice (kg)	Mass of melted ice (kg)
0.075						

of energy is the same as the amount transferred away from one kilogram of water when it freezes.)

If you have time, carry out Extension 1-2 to calculate the approximate *latent heat of fusion* for the ice to water phase transition using your data in Table 1-1.

Extension 1-2: Latent Heat of Fusion of Water

1. Calculate the total heat energy transferred to the ice/water mixture to melt the ice. Use the power rating of the heater and the total time of the heat pulses. Show your calculations below. (Remember, a watt is a joule/second.)

Power rating marked on the heater: _____W

Total heat energy transferred to melt the ice: _____ J

2. Calculate the latent heat of fusion of the ice. Use the amount of heat energy transferred that went to melting the ice and the mass of the ice to find the amount of heat energy needed to melt one kilogram of ice. Show your calculations.

Latent heat of fusion: _____ J/kg

Question E1-3: Compare the heat energy needed to melt one kilogram of ice with the heat energy needed to raise the temperature of one kilogram of water by one degree Celsius.

Question E1-4: Compare your value for the latent heat of fusion to the accepted value, 334×10^3 J/kg (or 334 J/g). Discuss the limitations in your experimental method that might account for any differences between these two values.

In the next investigation, you will examine another phase change. You will see what happens to the temperature of water when enough heat energy is transferred so that the water reaches its boiling-point temperature and begins to change to a gas (steam).

INVESTIGATION 2: CHANGE OF PHASE FROM LIQUID TO GAS

Prediction 2-1: On the axes that follow, sketch a graph of temperature vs. time for a cup of water initially at room temperature that is heated in the following way:

- Heat energy is transferred at a constant rate until the water reaches its boiling-point temperature. (Assume that this happens in the first 6 min.)

- Heat energy is transferred at the same constant rate as the water boils for the next 4 min.

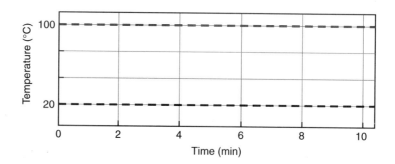

To test your prediction you will need

- computer-based laboratory system with one temperature sensor
- temperature software
- *RealTime Physics Heat and Thermodynamics* experiment configuration files
- heat pulser (relay box and 200-W immersion heater)
- room-temperature water
- container marked in mL
- stirring rod
- Styrofoam cup, 300 mL with tightly fitting cover
- vat (to prevent spills)
- electronic balance

Activity 2-1: Heating Water to Its Boiling Point: A Temperature History

1. Open the experiment file called **Water to Steam (L4A2-1)** to display the axes that follow. The heat pulser will also be set up to deliver 10-s pulses.

> **Reminder!!** Do not plug the heater into the pulser or press the HEAT button or key unless the heater is in the water.

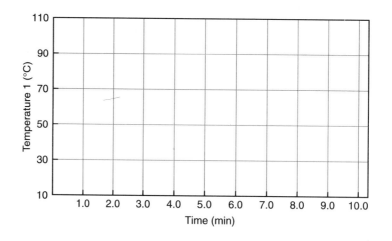

2. Accurately measure about 150 g of room-temperature water in the foam cup:

Mass of empty cup with cover: _____g

Mass of cup, cover, and water: _____g

Initial mass of water: _____g

Record in Table 2-1.

3. **Load the calibration file** for the temperature sensor, if it hasn't already been loaded.

4. Put the temperature sensor and heater in the cup. Let the heater sit on the bottom of the cup. Stir the water and record its initial temperature in Table 2-1.

5. **Begin graphing.** Also *begin stirring and keep stirring vigorously.* Start transferring heat pulses one after another, at a constant rate. *(As soon as one pulse stops, immediately press the **HEAT** button or key to begin the next pulse.)*

6. When the water begins to boil vigorously, record the temperature and time in the table and continue to pulse the heater. You may stop stirring, since the rolling boil will keep the water mixing.

Keep pulsing the heater until the water has been boiling for about 3 min.

7. As quickly as possible, remove the heater and temperature sensor and put the cover on the cup.

8. Record in Table 2-1 the number of heat pulses transferred to the water *after the water started to boil.*

9. Measure the final mass of the cup, cover, and water, calculate the mass of water converted to steam, and record in Table 2-1:

Final mass of cup, cover, and water: _____g

Mass of empty cup with cover: _____g

Final mass of water: _____g

Mass of steam produced: _____g

10. **Print the graph** and affix it over the previous axes. Indicate with an arrow where the water began boiling vigorously.

Question 2-1: Does your graph agree with the prediction you made? If not, describe the ways in which the observed behavior of boiling water was different than your prediction.

Table 2-1

Initial mass of water (kg)	Initial temperature (°C)	Time when boiling began (s)	Boiling point temperature (°C)	Number of heat pulses transferred after water was boiling	Total time of these heat pulses (s)	Final mass of water less steam (kg)	Mass of steam produced (kg)

Question 2-2: During the time that the temperature remained constant, what do you think happened to the heat energy you were transferring if it wasn't raising the temperature?

Comment: Most substances can exist in three states—solid, liquid, and gas. As you have seen, these *changes of state* or *phase changes* usually involve a transfer of heat energy. During a phase change, the substance can absorb heat energy without changing its temperature until the phase change is complete. The transferred energy increases the *internal energy* of the system, as you will explore further in the next investigation.

The amount of heat energy transfer required to transform one kilogram of water at its boiling point into steam is called the *latent heat of vaporization*. (This amount of energy is the same as the amount transferred away from one kilogram of steam when it condenses.)

If you have time, carry out Extension 2-2 to calculate the approximate *latent heat of vaporization* for the water to steam phase transition using your data from this activity.

Extension 2-2: Latent Heat of Vaporization of Water

1. Calculate the total heat energy transferred to the water *after it was boiling*. Use the power rating of the heater and the total time of the heat pulses from Table 2-1. Show your calculations below. (Remember, 1 W = 1 J/s.)

    Power rating marked on the heater: _____W

    Total heat energy transferred: _____ J

2. Calculate the latent heat of vaporization of the water. Use the amount of heat energy transferred after the water was boiling and the mass of steam produced from Table 2-1 to find the amount of heat energy needed to convert one kilogram of water to steam. Show your calculations.

    Latent heat of fusion: _____ J/kg

Question E2-3: Compare the heat energy needed to convert one kilogram of water to steam to that needed to convert one kilogram of ice to water. Why do you think that one is much larger than the other?

Question E2-4: Compare your value for the latent heat of vaporization to the accepted value, 2.26×10^6 J/kg (or 2260 J/g). Discuss the limitations in the experimental method you used that might account for any differences between these two values.

INVESTIGATION 3: WORK DONE BY A GAS, HEAT ENERGY TRANSFER, INTERNAL ENERGY, AND THE FIRST LAW OF THERMODYNAMICS

One system we will meet often in our study of thermodynamics is a mass of gas confined in a cylinder with a movable piston. The use of a gas-filled cylinder is not surprising since the development of thermodynamics in the eighteenth and nineteenth centuries was closely tied to the development of the steam engine, which employed hot steam confined in just such a cylinder.

In thermodynamics we are interested not only in heat energy and work, but in how the two interact. For example, if we transfer heat energy to a gas, can we get it to do work? In this investigation, you will begin with some qualitative observations to examine the concept of work done by the gas in a cylinder.

At your lab station you will find a number of syringes that are basically cylinders with movable pistons. By making some simple observations with these syringes, you can begin to appreciate how an expanding gas can do work.

You will need

- plastic syringe with the end sealed
- foam pad

Activity 3-1: Work Done by a Gas in a Cylinder

Try compressing the air in the syringe with the end sealed by pushing the piston down against the foam pad on the table. Then let it go, and see what happens.

Question 3-1: Does it take effort to compress the gas? Do you have to do work on the gas to compress it? (Did you apply a force over a distance?) What happens when you let go—Does the gas spring back?

In thermodynamics, pressure (defined as the component of force that is perpendicular to a given surface for a unit area of that surface) is often a more useful quantity than force alone, since it is independent of the cross-sectional area of the cylinder. It can be represented by the equation

$$P = \frac{F_\perp}{A}$$

In the next activity you will explore why pressure is more useful than force in describing the behavior of gases. You will also extend the definition of work developed earlier in the course and combine it with this definition of pressure to calculate the work done by a gas on its surroundings as it expands out against the piston with a (possibly changing) pressure P.

You will need

- two very different diameter low-friction glass syringes
- nonexpanding plastic tubing to connect the syringes
- vernier caliper
- ring stand and two clamps to hold the syringes

> Warning: Be very careful with the glass syringes. If they fall, they will break! In the following activity, take special care that the pistons do not fly out of the syringes.

Activity 3-2: Pressure and Work

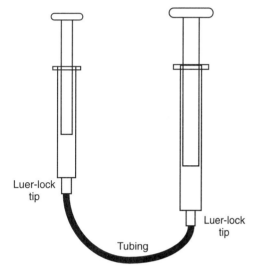

Luer-lock tip

Tubing

Luer-lock tip

1. Be sure that the pistons in the syringes are free to move. If sticky, remove the pistons and dip them in *distilled* water.

2. Mount the two syringes on the ring stand and connect using the tubing, leaving an air space in the small syringe of about half its length.

3. Push down on the smaller diameter syringe and observe what happens to the larger diameter one.

4. Push down on the larger diameter syringe and see what happens to the smaller diameter one.

5. Place a 100-g (0.1-kg) mass on top of the handle of the larger diameter syringe. Then find what mass must be placed on the handle of the smaller diameter syringe to just support the 100-g mass, and record it in Table 3-2.

6. Measure the diameters and masses of the pistons in the two syringes, and record the values in Table 3-2.

7. Calculate the cross-sectional area of each piston and record in the table.

8. Calculate the total weights of the pistons plus the supported masses, and record in the table.

9. Calculate the pressure in each syringe using the definition of pressure as force per unit area:

Pressure in smaller syringe: _____

Pressure in larger syringe: _____

Question 3-2: Based on your observations, does the force or the pressure applied to one syringe get transmitted to the other? Be quantitative.

Question 3-3: Why is pressure a more useful concept than force for a gas in a cylinder? Explain based on your observations.

Table 3-2

	Supported mass (kg)	Mass of piston (kg)	Total mass (kg)	Total weight (force) (N)	Diameter of piston (m)	Area of piston (m^2)
Larger syringe	0.10					
Smaller syringe						

Comment: This two-syringe combination works on the same principle as a hydraulic lift used to multiply forces. One application is the car lift at your auto mechanic's shop.

From your experiences with the syringes, do you expect an expanding gas inside a cylinder to do work? You have probably heard the definition of work in a lecture or seen it in your text. If a force F acts on an object and the object moves a distance Δx, the work is $\Delta W = F \Delta x$. Using this definition of work and the definition of pressure, you can show that the work done by a gas on its surroundings as it expands out against the piston with a (possibly changing) pressure P can be calculated from

$$\Delta W = P \Delta V$$

Gas at pressure P exerts a force on the piston given by $F = PA$

Piston of cross-sectional area A moves a distance Δx. $\Delta V = A \Delta x$

Question 3-4: Show that the above expression for ΔW follows from $\Delta W = F \Delta x$. (**Hint:** See the preceding diagram.)

Suppose you lift a ball of mass m up from the floor through a distance y. The change in the ball's potential energy is $\Delta E_{\mathrm{pot},g} = mgy$. The work done by you against the force of gravity is related to the change in the ball's potential energy so that $\Delta E_{\mathrm{pot},g} = -W_g$. (Keep in mind that the force of gravity on the ball and the displacement y are vectors.)

This relationship is true for *any* system where mechanical energy is conserved. By doing work against gravity, you are storing energy in the form of potential energy $E_{\mathrm{pot},g}$. But what about systems where mechanical energy is apparently *not* conserved?

Is it possible to generalize this relationship for some of these systems? The answer is yes, but we have to give a new meaning to our potential energy E_{pot}. In thermodynamics, E_{pot} is called the *internal energy, E_{int},* and represents *any* way of storing energy inside a system. The internal energy of a system is the sum of all sorts of energies, including the helter-skelter translational kinetic energies of molecules in a gas, the vibrational energies of gas molecules or atoms in a crystal, and the rotational energies of spinning gas molecules. One way to increase the internal energy of a system is to transfer heat energy to it as you did when you melted ice or produced steam.

Transferring heat energy to a system could serve to increase its internal energy, but it might result instead in the system doing work on its surroundings.

Prediction 3-1: What do you think would happen if you put a syringe with a sealed end in hot water? Would its piston experience a force? Could it do work on its surroundings?

To test your prediction you will need

- plastic syringe with the end sealed
- 0.5-L container and hot water (at least 80°C)

Activity 3-3: The Heated Syringe

1. Take the syringe with a sealed end and, *while holding the piston fixed in place*, submerge most of the syringe in very hot water.

2. After a minute or so, release the piston and let it move freely.

Question 3-5: What do you think might have happened to the gas while you heated it but prevented it from changing its volume by holding the piston fixed?

Question 3-6: What happened when you released the piston and let the gas expand? Did the gas do any work? Is this what you predicted would happen in Prediction 3-1?

You should have concluded from the last activity that the transfer of heat energy to a system can either cause it to do work on its surroundings or increase its internal energy. What is the relationship between heat energy transfer, changes in a system's internal energy, and the work done by the system? We picture E_{int} as the "true" energy of the system. In the theory of thermodynamics, E_{int} is called a "state" variable, a quantity that tells us some things we need to know to calculate useful things about a system, such as its temperature. To help us understand how E_{int} is related to work and heat energy, we will take a closer look at what happened to the gas confined in the syringe in Activity 3-3.

Suppose that the piston of the syringe is clamped in place while the syringe is immersed in hot water. The piston can't move, so no work can be done. However, since the water is initially at a higher temperature than the gas in the syringe, we expect that heat energy is transferred from the water to the gas. This causes the temperature of the gas to increase and the temperature of the water to decrease. The heat energy transfer can be calculated using the equation $Q = cm\,\Delta T$, where c is the specific heat and ΔT is the temperature change of the water.

Assuming that the system is insulated so that no heat energy can be transferred to the surroundings, the transferred heat energy Q must equal the increase in internal energy of the gas. This is based on a belief that energy is conserved in the interaction between the hot water and the gas.

Suppose instead that we release the piston and allow the gas to do work as it expands against the piston. We could calculate the amount of work the expanding gas did by evaluating $\Delta W = P\,\Delta V$ for the whole process. Where did the energy to do this work come from? The only possible source is the internal energy of the gas, which must have decreased by an amount W. The total change in the internal energy of the air trapped in the syringe must be given by

$$\Delta E_{int} = Q - W$$

This relationship between absorbed heat energy, work done on the surroundings, and the change in internal energy is believed to hold for any system, not just a syringe filled with gas. It is known as the *first law of thermodynamics.*

The first law of thermodynamics has been developed by physicists based on a set of very powerful inferences about forms of energy and their transformations. We ask you to try to accept it on faith. The concepts of work, heat energy transfer, and internal energy are subtle and complex. For example, *work* is not simply the motion of the center of mass of a rigid object or the movement of a person in the context of the first law. Instead, we have to learn to draw system boundaries and total the mechanical work done by the system inside a boundary on its surroundings outside the boundary.

The first law of thermodynamics is a very general statement of conservation of energy for thermal systems. It is not easy to verify in an introductory physics laboratory, and it is not derivable from Newton's laws. Instead, it is an independent assertion about the nature of the physical world.

There are many ways to achieve the same internal energy change, ΔE_{int}. To achieve a small change in the internal energy of gas in a syringe, you could transfer a large amount of heat energy to it and then allow the gas to do work on its surroundings. Alternatively, you could transfer a small amount of heat energy to the gas and not let it do any work at all. The change in internal energy, ΔE_{int}, could be the same in both processes. ΔE_{int} depends only on $Q - W$ and not on Q or W alone.

Question 3-7: Can you think of any situations where W is negligible and $\Delta E_{int} = Q$? (**Hint:** Is it necessary to do work on a cup of hot coffee to cool it? Can you think of similar situations?)

Question 3-8: How could you arrange a situation where Q is negligible and in which $\Delta E_{int} = -W$? Such situations have a special name in thermodynamics. They are called *adiabatic processes*. (*Adiabatic* means with no heat energy transferred into or out of the system.)

HOMEWORK FOR LAB 4:
THE FIRST LAW OF THERMODYNAMICS

1. Describe what happens to the temperature of water between 0 and 100°C when heat energy is transferred to it at a constant rate.

2. Describe what happens to the temperature of a water–ice mixture originally at 0°C when heat energy is transferred to it at a constant rate. Sketch a temperature history on the axes below. Indicate on your graph where the ice has completely melted.

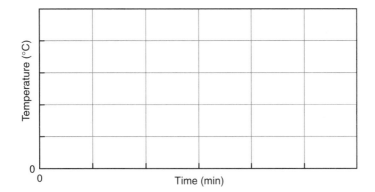

3. Explain what happens to the heat energy transferred to the ice and water mixture if the temperature does not rise while the ice is melting. How does the internal energy of the ice and water mixture change during this process? Does it increase, decrease, or remain the same? Explain.

4. Suppose you start with 250 g of ice at 0°C. Calculate the amount of heat energy that must be transferred to the ice to melt it. (Use 334×10^3 J/kg for the latent heat of fusion of ice.) Show your calculations.

5. A mixture of 150 g of ice and 300 g of water is at 0°C. How many joules of heat energy must be transferred to bring this mixture to a final temperature of 75°C? Assume that the heat energy transferred to the room is very small.

Show your calculations. [Use 334×10^3 J/kg (334 J/g) for the latent heat of fusion of ice and 4190 J/kg°C (4.19 J/g°C) for the specific heat of water.]

6. Describe what would happen to the water in Question 2 if you continued to transfer heat energy at a constant rate even after the ice has melted. Sketch the temperature history on the axes below. Indicate on your graph the time when the water begins to boil.

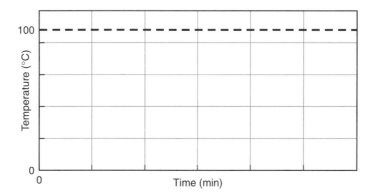

7. Explain what happens to the heat energy transferred to the water and steam mixture if the temperature does not rise while the water is boiling. How does the internal energy of the water and steam mixture change during this process? Does it increase, decrease, or remain the same? Explain.

8. Suppose you start with 250 g of ice at 0°C. Calculate the amount of heat energy that must be transferred to convert the ice to steam at 100°C. Show your calculations. [Use 334×10^3 J/kg (334 J/g) for the latent heat of fusion, 2.26×10^6 J/kg (2260 J/g) for the latent heat of vaporization, and 4190 J/kg°C (4.19 J/g°C) for the specific heat of water.]

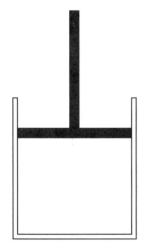

9. Gas is in a cylinder sealed with a piston. If you push down quickly on the piston and compress the gas, is work done on or by the gas? What happens to the internal energy of the gas? What happens to the temperature of the gas?

10. Now you transfer heat energy to the gas in the cylinder, but hold the piston so that it cannot move. Is work done on or by the gas? What happens to the internal energy of the gas? What happens to the temperature of the gas?

11. After holding the piston in Question 10 for a short while, you let it go. What happens to the piston? Is work done on or by the gas? What happens to the internal energy of the gas? What happens to the temperature of the gas?

Name_____   Date_____

Pre-Lab Preparation Sheet for Lab 5:
The Ideal Gas Law

(Due at the beginning of Lab 5)

Directions:
Read over Lab 5 and then answer the following questions about the procedures.

1. What is your Prediction 2-1?

2. Why must the temperature be kept constant in Activity 2-1?

3. What is kept constant in Activity 2-2? How is this done?

4. What is kept constant in Activity 2-3? How is this done?

5. What is your Prediction 2-3?

LAB 5:
THE IDEAL GAS LAW

... the hypothesis, that supposes the pressures and expansions to be in reciprocal proportions ...

—Robert Boyle

OBJECTIVES

- To understand how a gaseous system may be characterized by temperature, pressure, and volume.

- To examine the relationship between any two of these variables when the third is kept constant.

- To understand and be able to use the ideal gas law, that describes the relationship between pressure, volume and temperature.

- To understand pressure in a gas based on an idealized model of the microscopic motions of molecules, and use this model as a possible explanation for the ideal gas law.

OVERVIEW

In introductory physics we often talk about matter as if it were continuous. We don't need to invent aluminum atoms to understand how a ball rolls down a track. As early as the fifth century B.C., Greek philosophers, such as Leucippus and Democritus, proposed the idea of "atomism." They pictured a universe in which everything is made up of tiny "eternal" and "incorruptible" particles, separated by "a void." Today, we think of these particles as atoms and molecules.

In terms of every day experience, molecules and atoms are hypothetical entities. In just the past forty years or so, scientists have been able to "see" molecules using electron microscopes and field ion microscopes. But long before atoms and molecules could be "seen," nineteenth-century scientists, such as James Clerk Maxwell and Ludwig Boltzmann in Europe and Josiah Willard Gibbs in the United States, used these imaginary, small-scale *microscopic* entities to construct models that account for the large-scale *macroscopic* properties of thermodynamic systems.

Even a small container filled with a gas contains a very large number of molecules, on the order of 10^{23}! Since it is impossible to use Newton's laws of motion to keep track of what each of these molecules is doing at any moment, we must characterize the behavior of a gas by the macroscopic quantities: volume, V; pressure, P; and temperature, T. *Kinetic theory* is the area of physics that uses Newton's laws and averages of molecular behavior to explain the relationship between P, V, and T.

In this lab, you will first look at how pressure is measured. After this, your goal will be to determine experimentally how P, V, and T are related. You will carry out experiments to relate P and V, P and T, and V and T. Finally, you will combine the relationships you discover into the *ideal gas law*.

In the last investigation, you will examine some simulations of molecular motions to develop a microscopic model that can explain how gases can exert pressure. These simulations will also help you begin to understand the relationship between P, V, and T.

INVESTIGATION 1: MEASURING PRESSURE

Before undertaking the study of the gas laws, we need to learn something about how pressure is measured. We will use two devices, a manometer and a computer-based pressure sensor.

To examine the operation of a manometer, you will need

- 2 translucent soda straws, about 20 cm long

- length of Tygon tubing, about 20 cm long

- small balloon

- ring stand with two clamps to hold straws

Activity 1-1: Measuring Pressure With a Manometer

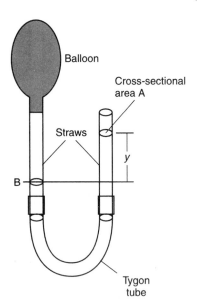

Balloon

Cross-sectional area A

Straws

y

B

Tygon tube

At your lab station there should be a manometer that looks like the one pictured on the left. The column of air above each unit area of the earth's surface exerts a force. This is atmospheric pressure, which is often denoted P_{atm}. The columns of water on each side of the manometer experience this atmospheric pressure.

If one column is forced a distance y higher than the other, there is an additional volume of liquid $\Delta V = Ay$, where A is cross-sectional area of the column of water. Thus, an additional weight $mg = \rho \, \Delta Vg = \rho g Ay$ of water must be supported, where ρ is the density of water. Since pressure is the force per unit area, the *additional* pressure ΔP at level B is given by $\Delta P = mg/A = \rho gy$. Therefore, the pressure in the left straw must be given by

$$P = P_{atm} + \rho gy$$

By measuring y, you can determine the pressure:

1. Put some water in the bottom of the manometer tubes.

2. Blow a small amount of air into the balloon and attach the balloon over the end of the left straw.

3. Measure the difference in heights of the two columns.

Question 1-1: Calculate the pressure of gas in the balloon using $\rho = 1000$ kg/m³ and $g = 9.80$ m/s². (**Notes:** 1 Pa (pascal) $= 1$ N/m². P_{atm} varies depending on the altitude where this activity is carried out.)

An easier and more accurate way to measure pressure is to use a computer-based pressure sensor. To examine the operation of this system, you will need

- computer-based laboratory system with one pressure sensor
- data logger software
- *RealTime Physics Heat and Thermodynamics* experiment configuration files

Activity 1-2: Using a Computer-Based Pressure Sensor

1. Plug the pressure sensor into the interface and open the data logger software.

2. Open the experiment file called **Measuring Pressure (L5A1-2)** to display the pressure digitally on the screen.

3. **Load the calibration file** for the pressure sensor to read pressures in Pa.

4. Record the pressure with the sensor open to the air.

5. Blow up the balloon and attach it to the pressure sensor. Record the pressure in the balloon.

Question 1-2: Was the pressure in the filled balloon significantly different from atmospheric pressure? If so, why is the pressure different?

INVESTIGATION 2: BEHAVIOR OF A GAS IN TERMS OF *P, V,* AND *T*

How do the three variables, pressure P, volume V, and temperature T, of a gas depend on each other? Now that you know how to measure pressure and temperature with computer-based pressure and temperature sensors, you can explore the relationships between these quantities. To simplify this investigation, you will look at the behavior of any two of these variables, while the third is kept constant.

Comment: In SI units, pressure is measured in Pa (N/m²) and volume is measured in m³. Temperature can be measured in °C (degrees Celsius) or in K (kelvins). It turns out that the relationships that involve temperature appear simpler if temperature is measured in K, the Kelvin or absolute temperature scale. The reason for this can be traced back to the definition of temperature and its meaning on a microscopic scale. We will talk more about this later. For now, you can simply set up your software to display the readings of your temperature sensor in K.

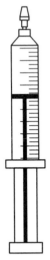

You will start by examining the relationship between pressure P and volume V by doing measurements on the air in a syringe. Since the syringe is in thermal contact with the surrounding air, if you change the volume of the gas relatively slowly, the gas in the syringe remains in thermal equilibrium with the surroundings. Another way of referring to a process that takes place at constant temperature is to call it *isothermal*. First make a prediction.

Prediction 2-1: As you compress the air in a syringe by pushing the piston in *slowly*, what will happen to the pressure? What do you think will be the mathematical relationship between pressure P and volume V?

To test your prediction you will need

* 20-mL plastic syringe (with the needle removed)
* short piece of Tygon tubing (to attach syringe to pressure sensor)
* computer-based laboratory system with one pressure sensor
* data logger software
* *RealTime Physics Heat and Thermodynamics* experiment configuration files

Activity 2-1: Isothermal Volume Change for a Gas

The approach to obtaining measurements is to trap a volume of air in the syringe and then compress the air slowly to smaller and smaller volumes by pushing in the piston. The gas should be compressed slowly so it will always have time to come into thermal equilibrium with the room (and thus be at room temperature). You should take pressure data for about 5 different volumes.

1. Attach the end of an unsealed syringe to the pressure sensor using the Tygon tubing. Start with the piston at 20 mL. Be sure that the valve on the pressure sensor is completely closed to the atmosphere.

2. Open the experiment file called **Pressure vs. Volume (L5A2-1)** to display the axes that follow. This will also set up the software in **prompted event mode** so that you can continuously measure pressure and decide when you want to **keep** a value. Then you can **enter** the measured volume.

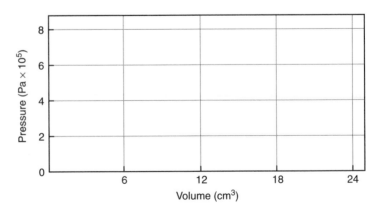

3. **Load the calibration file** for the pressure sensor.

4. Enter the volume of the pressure sensor given to you in the second column of Table 2-1.

5. Estimate the volume of the tubing between the syringe and the sensor from its inside diameter (which will be given to you) and its length. Show your calculation.

Inside diameter:_____cm   Length:_____cm

Estimated volume of tubing:_____cm^3

Enter this value in Table 2-1, and use this table to correct the volumes measured with the syringe.

6. As you squeeze down on the piston slowly, the computer will display the pressure. When the pressure reading is stable, you can **keep that value** and then **enter the total volume** of air from Table 2-1.

7. Repeat this for five different volumes of the syringe.

8. Use the **fit routine** to find a relationship between P and V.

9. **Print the graph** and affix it over the previous axes.

Question 2-1: What is the relationship between P and V? Is it proportional, linear, inversely proportional, or something else? Did this agree with your prediction?

Question 2-2: Write down the relationship between the initial pressure and volume (P_i, V_i) and the final pressure and volume (P_f, V_f) for an isothermal (constant-temperature) process.

The relationship that you have been examining between P and V for a gas with the temperature and amount of gas held constant is known as Boyle's law.

Table 2-1

Volume of air in syringe (cm^3)	Volume of sensor (cm^3)	Estimated volume of tubing (cm^3)	Total volume of air in system (cm^3)

How does a change in temperature affect a gas? You can explore the effect of temperature on pressure with the volume kept constant and on volume with the pressure kept constant. Let's start with P vs. T. A process in which the volume is kept constant is called *isovolumetric* or *isochoric*.

Prediction 2-2: A glass flask of fixed volume containing air is moved from a water bath at one temperature to one at a higher temperature. (The flask is left long enough so that the air is in thermal equilibrium with the water baths.) How does this affect the pressure inside the flask? What do you think is the mathematical relationship between P and T with the volume held constant?

To test your prediction you will need

- 25-mL glass boiling flask with one-hole stopper and glass tube
- short piece of Tygon tubing (to attach flask to pressure sensor)
- large insulated container (e.g., Styrofoam cup)
- hot water (about 70°C)
- ice
- computer-based laboratory system with one pressure sensor and one temperature sensor
- data logger software
- *RealTime Physics Heat and Thermodynamics* experiment configuration files

Activity 2-2: Isovolumetric Temperature Change for a Gas

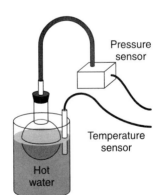

1. Connect the flask to the pressure probe. The flask contains your test volume of gas (air).
2. Connect the pressure and temperature sensors to the computer interface.
3. Open the experiment file called **Measuring P and T (L5A2-2)** to display the axes that follow. The software will be set up in **event mode** in which you can observe the values being measured and decide when you want to **keep** them.
4. **Load the calibration files** for the pressure and temperature sensors.
5. **Begin graphing** and place the flask in the cup containing water at about 70°C (343 K). Keep the flask submerged. When the pressure and temperature stop changing, **keep the data point.**
6. Add a small amount of ice to the hot water. Stir until the temperature and pressure stop changing and then **keep the data point.**
7. Repeat step 6 for at least four more different temperatures, down to about 10°C (283 K).
8. Use the **fit routine** to find a relationship between P and T.
9. **Print the graph** and affix it over the axes that follow.

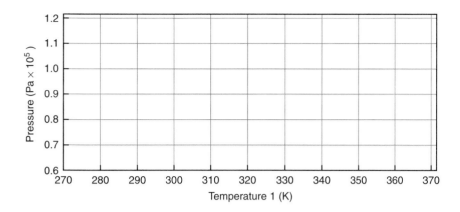

Question 2-3: What is the relationship between P and T? Is it proportional, linear, inversely proportional, or something else? Did this agree with your prediction?

Question 2-4: Write down the relationship between the initial pressure and temperature (P_i, T_i) and the final pressure and volume (P_f, T_f) for an isovolumetric process.

The relationship that you have been examining between P and T for a gas with the volume and amount of gas held constant is known as Gay–Lussac's law.

How does a change in temperature affect the volume of a gas when the pressure is kept constant? A process in which the pressure is kept constant is called *isobaric*.

Prediction 2-3: A low-friction syringe containing air is moved from a water bath at one temperature to one at a higher temperature. (The syringe is left long enough so that the air is in thermal equilibrium with the water baths.) How does the volume of the air change? What do you think will be the mathematical relationship between V and T with the pressure held constant?

To test your prediction you will need

- 20-mL plastic syringe
- 25-mL glass boiling flask with a one-hole stopper and T-connector
- short pieces of Tygon tubing (to attach flask to pressure sensor and to syringe)
- large insulated container (e.g., Styrofoam cup)
- hot water (about 70°C)
- ice
- computer-based laboratory system with one pressure sensor and one temperature sensor
- data logger software
- *RealTime Physics Heat and Thermodynamics* experiment configuration files

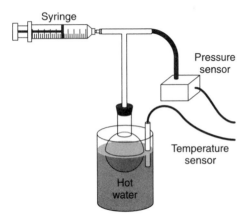

Syringe

Pressure
sensor

Temperature
sensor

Hot
water

Activity 2-3: Isobaric Temperature Change for a Gas

1. Connect the flask to the pressure probe and to the syringe.

2. Connect the pressure and temperature sensors to the computer interface.

3. Open the experiment file called **Measuring V and T (L5A2-3)** to display the axes that follow. This will also set up the software in **prompted event mode** so that you can continuously measure temperature and decide when you want to **keep** a value. Then you can **enter** the measured volume.

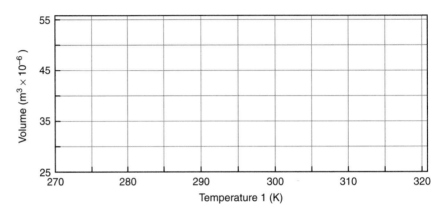

4. **Load the calibration files** for the pressure and temperature sensors.

5. Enter the volume of the pressure sensor given to you in the third column of Table 2-3.

6. Estimate the volume of the tubing between the syringe and flask and the sensor from its inside diameter (which will be given to you) and its length. Show your calculation.

Inside diameter:_____cm   Length:_____cm

Estimated volume of tubing:_____cm^3

Enter this value in the fourth column of Table 2-3, and use this table to calculate the total volumes of air.

Table 2-3

Volume of air in syringe (cm^3)	Volume of flask (cm^3)	Volume of sensor (cm^3)	Volume of tubing (cm^3)	Total volume of air (cm^3)
	25			
	25			
	25			
	25			
	25			

7. **Begin graphing** and place the flask in the cup containing water at about 70°C (343 K). Keep the flask submerged. When the pressure and temperature stop changing, open the valve and pull out the piston on the syringe to 20 mL. Close the valve and keep it closed for the remainder of this activity.

8. When the pressure and temperature become stable, **keep the data point. Enter the total volume** from Table 2-3.

9. Record the pressure: _____

10. Add a small amount of ice to the hot water. Adjust the syringe so that when the temperature and pressure stop changing, the pressure is the same as before. (This assures that the process is *isobaric,* i.e., pressure is being kept constant.) **Keep the data point** and **enter in the new total volume** from Table 2-3.

11. Repeat this procedure in at least five steps until you reach the minimum possible volume.

12. Use the **fit routine** to find the relationship between *V* and *T*.

13. **Print the graph** and affix it over the previous axes.

Question 2-5: What is the relationship between *V* and *T*? Is it proportional, linear, inversely proportional, or something else? Did this agree with your prediction?

Question 2-6: Write down the relationship between the initial volume and temperature (V_i, T_i) and the final volume and temperature (V_f, T_f) for an isobaric process.

The relationship that you have been examining between *V* and *T* for a gas with the pressure and amount of gas held constant is known as Charles' law.

We have seen how pressure depends on temperature at constant volume, how volume depends on temperature at constant pressure, and how pressure and volume are related at a constant temperature.

You have probably heard of the ideal gas law:

$$PV = (\text{constant})T$$

Question 2-7: Are the relationships you found in the past three activities consistent with the ideal gas law? Explain based on your investigations of Boyle's law, Gay–Lussac's law, and Charles' law.

INVESTIGATION 3: PRESSURE AND BEHAVIOR OF MOLECULES IN A GAS

If you have more time, work on the following extensions, which will give you some idea of the origin of pressure on a microscopic level.

Some of our ancestors believed in the reality of witches. In fact, they thought that they had good evidence that witches existed, good enough evidence to accuse some people of being witches. We believe in atoms. Are we truly more scientific than they were?

Question E3-1:   Describe several reasons why you do or do not believe that matter consists of atoms and molecules, even though you have never seen them with your own eyes.

Question E3-2:   What happens when heat energy is transferred into a substance? If you believe that substances are made of atoms and molecules, how might you use a belief in their existence to explain the change in volume (under constant pressure) of a heated gas as you have seen in the previous lab?

Extension 3-1: Origin of Pressure

Is it possible to describe the behavior of a gas as a collection of moving molecules? To answer this question, let's first use a simulation to observe the pressure exerted by a hypothetical molecule undergoing elastic collisions with the walls of a two-dimensional box. Then we will use another simulation to look at what happens to the molecules in a container when the gas is compressed. To make these observations you will need

- Single Particle Movie (L5E3-1)
- Syringe Movie (L5E3-2)
- movie viewer software

1.   Open the movie viewer software and view the **Single Particle Movie (L5E3-1)** simulation of the single gas molecule in a box several times.

2.   Observe what happens to the x component of the velocity when the molecule collides with a side wall.

3.   Observe what happens to the y component of the velocity when the molecule collides with the top or bottom wall.

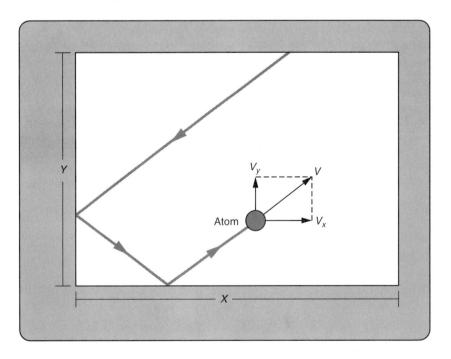

Question E3-3: What happens to the velocity components when a collision with a wall takes place? What is the change in the component of momentum perpendicular to the wall, $p_x = mv_x$ or $p_y = mv_y$, produced by a collision?

Question E3-4: Based on your observations, does the wall exert a force on the molecule during a collision? How is this related to the rate of change of momentum of the molecule? Does the molecule exert a force on the wall?

Question E3-5: What would happen if there were a very large number of molecules bouncing around in the container instead of just one molecule?

Question E3-6: Can you now describe the origin of pressure in a gas? Explain.

Extension 3-2: Pressure vs. Volume

Prediction E3-1: What would happen to the number of collisions per unit time with the walls of the container if one of the walls started to move inward making the volume smaller? What would happen to the momentum transferred to a molecule in each collision with this moving wall? What would happen to the pressure in the gas?

1. View the **Syringe Movie (L5E3-2)** simulation of several gas particles in a syringe.

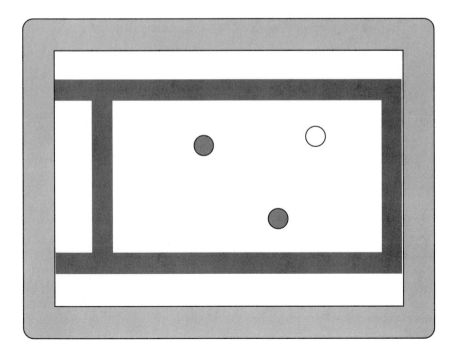

2. Observe what happens to the number of collisions per unit time with the walls as the piston is pushed in.

Question E3-7: What happens to the number of collisions per unit time with the piston as it is pushed in? What happens to the rate of change of momentum of the particles in these collisions? What should happen to the pressure (force per unit area)?

Question E3-8: How do your answers to Question E3-7 compare to your observations of Boyle's law in Activity 2-1? Explain.

These very simple kinetic theory simulations may help you begin to see what is going on inside a gas and to understand how pressure and volume might be related. Try to imagine a gas made up of a very large number of molecules, constantly bouncing around off the walls (and off each other). The rate of change of momentum for all of the molecules colliding with a wall is equal to the force exerted on the wall, and the force per unit area is the pressure.

One important difference from the simulations is that the large number of molecules in a real gas do not all have the same velocity. However, the pressure can still be related to an average velocity of the molecules.

The simulations have told us nothing about the temperature of the gas. It is possible from kinetic theory to show that temperature is a measure of the average kinetic energy of the molecules in a gas as they move around randomly colliding with each other and with the walls. The higher the temperature of a gas, the larger the average kinetic energy of the molecules. The larger the average kinetic energy, the larger the average velocity of a molecule. This has two consequences: (1) The average momentum carried by a molecule is larger, and (2) the molecules make more collisions with the walls per unit time.

Question E3-9: Based on your simple kinetic theory observations and this knowledge of what temperature measures, can you explain why the pressure should vary with temperature in the way you observed Activity 2-2?

Name_____ Date_____ Partners_____

HOMEWORK FOR LAB 5:
THE IDEAL GAS LAW

1. Explain the operation of a manometer. Why is the measured pressure related to the difference in heights of the two columns of water?

2. Describe the evidence from this experiment which supports Boyle's law, PV = constant for an isothermal process. Describe any experimental difficulties you had in exploring this law.

3. Describe the evidence from this experiment which supports Gay–Lussac's law, P/T = constant for an isovolumetric process. Describe any experimental difficulties you had in exploring this law.

4. Describe the evidence from this experiment which supports Charles' law, V/T = constant for an isobaric process. Describe any experimental difficulties you had in exploring this law.

5. Which of the three laws was the most difficult to explore experimentally? Why?

6. Why was it necessary to keep one of the variables P, V, or T constant in each experiment?

7. The ideal gas law is given as $PV = nRT$, where n is the number of moles of gas (1 mole $= 6.02 \times 10^{23}$ molecules) and R is the universal gas constant ($R = 8.31$ J/mol-K). Explain why the product of pressure and volume is constant in an *isothermal* process.

Explain why the ratio of pressure and temperature is constant in an *isovolumetric* process.

8. Explain the origin of pressure of a gas in a container based on the motions of molecules in the gas, their average momentum, and their collisions with the walls of the container.

9. Based on the kinetic theory definition of temperature as the average kinetic energy of a molecule in an ideal gas and your answer to Question 8, explain why the pressure of a gas increases when it is compressed while kept at a constant temperature.

10. Based on the kinetic theory definition of temperature as the average kinetic energy of a molecule in an ideal gas and your answer to Question 8, explain why the pressure of a gas increases when its temperature is increased while its volume is kept constant.

Name_____ Date_____

PRE-LAB PREPARATION SHEET FOR LAB 6:
HEAT ENGINES

(Due at the beginning of Lab 6)

Directions:
Read over Lab 6 and then answer the following questions about the procedures.

1. What is the purpose of the heat gun or hair dryer in Activity 1-1?

2. How is work calculated from a *P-V* diagram? How about for a closed cycle?

3. Describe the parts of the "incredible mass-lifting heat engine."

4. What is an *adiabatic* process? Which parts of the mass-lifting heat engine cycle are approximately *adiabatic*?

5. How will the work done by the mass-lifting heat engine in a cycle be calculated?

LAB 6: HEAT ENGINES

The production of motion in a steam engine is always accompanied by a circumstance which we should particularly notice. This circumstance is the passage of caloric from one body where the temperature is . . . elevated to another where it is lower.

—S.N.L. Carnot

OBJECTIVES

- To be able to describe a heat engine in terms of an energy flow diagram and to calculate the work done in a cycle.

- To examine the relationship between temperature and volume for both adiabatic and isothermal expansions and compressions of an ideal gas.

- To investigate, both theoretically and experimentally, the relationship between work done by a heat engine and changes in the pressure and volume of the engine's working medium.

- To examine the efficiency of a heat engine in converting heat energy input to useful work output.

OVERVIEW

In general, engines convert various forms of energy into mechanical work. For example, an athlete uses chemical energy released during the oxidation of molecules obtained from food to do mechanical work. The efficiency of an athlete's muscles in transforming chemical energy into work is at best only about 20%. The other 80% of the chemical energy released during physical activity is ultimately converted into waste heat energy. Since an athlete must maintain her internal body temperature at about 37°C, this waste heat energy must be transferred to her surroundings.

The nineteenth-century industrial revolution was based on the invention of heat engines. Heat engines have much in common with the chemical engines that power humans. For instance, both human engines and heat engines extract heat energy at a higher temperature, do work, and then transfer waste heat energy to lower temperature surroundings. Even though the 20% efficiency of a human engine seems low, it is far higher than the efficiency of a heat engine working between the same two temperatures. When you complete this unit, you will have a better understanding of how the laws of thermodynamics allow us to place absolute limits on the efficiency of heat engines.

The steam engine, which ushered in the industrial revolution, and the internal combustion engine both depend on *cycles* in which gases (the working medium of the engine) are alternately expanded and then compressed. The end result of these cycles is that a portion of heat energy transferred to a gas is converted into work. An understanding of the detailed physics of the expansion and compression of gases has helped engineers to design more efficient engines.

In this lab we will first look at the more general idea of a heat engine cycle that can convert transferred heat energy to work, using a simple rubber band engine as our example. We will then extend our discussion of the expansion and compression of gases from Labs 4 and 5, looking at a special type of compression called an *adiabatic* compression, during which there is no heat energy transfer from the gas. Next, we will examine a real heat engine that operates on expansions and compressions of a gas, and attempt to determine the amount of work that it will produce on the basis of our understanding of the physics of gas expansion and compression. Finally, we will calculate the efficiency of this engine and compare it to the maximum possible efficiency for an engine operating between the same temperatures.

INVESTIGATION 1: HEAT ENGINES AND CYCLES

Overview of Heat Engines

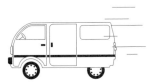

Internal combustion engines inside cars and trucks are examples of heat engines that burn a gasoline–air mixture in the cylinders. The railroad steam engine, which spanned the American continent in the mid-nineteenth century, is another example of a heat engine. The word "engine" conjures up an image of something that we start up, that runs, and that provides a continuous flow of work. In the case of a car, the work done by the engine accelerates us until we reach a suitable speed and then helps us maintain that speed by overcoming friction and air resistance.

The basic goal of any heat engine is to convert heat energy into work as efficiently as possible. This is done by taking a *working medium*—some substance that can expand (or contract) and thus do work when heat energy is transferred to it—and placing it in a system designed to produce work in continuous, repeated cycles.

An engine that is 100% efficient would have a working medium that transforms all of the heat energy transferred to it to useful work. Such a process would not violate the first law of thermodynamics, since energy would be converted from one form to another, and conserved. However, no heat engine has ever been capable of transforming *all* of the heat energy transferred to it into useful work. Some of the heat energy transferred in is always transferred back to the engine's surroundings at a lower temperature as waste heat energy. The universal existence of waste heat energy has led scientists to formulate the *second law of thermodynamics*. A common statement of the second law is simply that *it is impossible to transform all of the heat energy transferred to a system into useful work.*

When gasoline is burned inside an internal combustion engine, heat energy is transferred to the working medium—air in cylinders inside the engine.

In this lab you will explore the actual behavior of several simple heat engines. We can use any substance that changes its volume when heat energy is transferred to it as the working medium for a heat engine. The working medium is capable of transforming a fraction of the heat energy transferred into it to useful work. You will first explore the characteristics of a heat engine using a rubber band as the working medium. In the activities that follow you will study the compression and expansion of gases used as working substances in heat engines. As you complete this lab, we hope you will begin to understand that there are general principles that govern the operation of heat engines that do not depend on the detailed nature of the *working medium*.

Activity 1-1: The Rubber Band as a Heat Engine Medium

Let's examine what happens when a large rubber band with a weight at one end is heated with a heat gun or hair dryer. The equipment for this demonstration is as follows:

- large rubber band, 5 cm × 10 cm

- mass pan

- two 10-kg masses

- table clamp with long rod, short support rod, and right-angle clamp

- heat gun (or 1500-W hair dryer with Styrofoam housing around rubber band to keep heat energy from being transferred to the surroundings)

Prediction 1-1: What do you predict will happen when the rubber band that is stretched by a hanging weight is heated? Explain.

The rubber band engine will be demonstrated to you.

Question 1-1: Describe what actually happened when the rubber band was heated. Compare this to your prediction.

This behavior of a heated rubber band is helpful if all we want to do is to lift a weight once, but it hardly fits our intuitive notion of an engine. Let's say you own a factory that produces a canned beverage. (We will let you choose the beverage!) You wish to design a machine that lifts cans from the conveyor belt leaving the filling machine up to the conveyor belt entering the packing machine. You have at your disposal a large rubber band and a hair dryer. Can you design a mass lifter? It might look like the one in the following diagram.

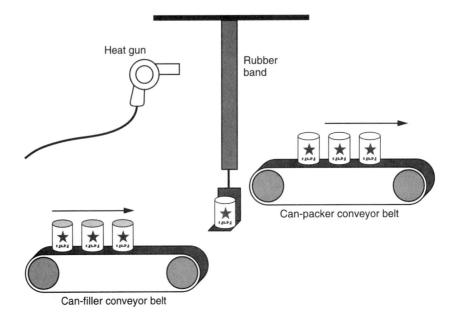

Question 1-2: What do you need to do to lift the first can from the level of the can-filler conveyor belt to the level of the can-packer conveyor belt?

Question 1-3: What do you need to do to the can once it has reached the packer conveyor belt?

Question 1-4: What must happen so that the lifter will be ready to pick up another can at the filler conveyor belt level? What has to happen to the rubber band?

Question 1-5: Can you use your answers above to describe a complete *cycle* that would repeatedly lift cans to the packer conveyor? Describe the steps in your cycle.

Question 1-6: Carefully point out where heat energy is transferred in your cycle, the direction of heat energy transfer, where work is done, and by which part of the engine that work is done.

Question 1-7: Would the engine work on a very hot day when the temperature inside the factory was as hot as the heated air coming from the hair dryer or heat gun?

Question 1-8: Does it appear likely that our rubber band lifter converts *all* of the heat energy transferred to it from the hair dryer into useful mechanical work (i.e.,

REALTIME PHYSICS: HEAT AND THERMODYNAMICS

work done lifting cans)? Does any of that transferred heat energy have to go else-where? (**Hint:** What has to happen to the rubber band after the lifted can is taken away but before it can pick up a new can from the lower belt?)

INVESTIGATION 2: ADIABATIC COMPRESSION OF GASES

It is important to understand what happens to gases when they undergo volume changes, since many practical devices, from automobile engines to refrigerators, use expanding gases to operate. If you have time, you should carry out Extension 2-1. Otherwise go on to Activity 2-2.

Extension 2-1: The Fire Syringe and the Rapid, Adiabatic Compression of Air

In Lab 5 we explored the behavior of a gas, and established that $PV = $ (constant)T describes the relationship between pressure P, volume V, and absolute (Kelvin) temperature T fairly well. Suppose that the volume of a gas is changed, i.e., the gas expands or is compressed. Is the ideal gas law by itself enough to tell us what happens to the temperature of the gas in this process?

Prediction E2-1: Suppose that you are told that the volume of a gas increases or decreases. Can the ideal gas law be used to calculate the change in temperature of the gas or do you need any additional information? Are there different ways in which an ideal gas can change its volume?

In Lab 5, you examined Boyle's law, the relationship between P and V in processes where T is kept constant—*isothermal* processes. In such a process the volume changes but the *temperature remains constant*. You accomplished this by changing the volume of the gas in a syringe *very slowly* so that the temperature of the gas had time to remain in thermal equilibrium with its surroundings—the air in the lab. There was enough time for heat energy to be transferred to or from the surroundings during the process.

But suppose that you compressed the gas very quickly or that the gas was in a cylinder that was well insulated from its surroundings. Then no heat energy is transferred to or from the gas during the process. Such a process is said to be *adiabatic*, which simply means "with no transfer of heat energy."

Prediction E2-2: If a gas is compressed adiabatically, what should happen to the temperature of the gas? (**Hint:** Consider the first law of thermodynamics, $\Delta E_{int} = Q - W$, and the relationship between internal energy E_{int} and T.)

To test your prediction, you will need

- fire syringe
- ruler
- glass thermometer

- tiny piece of tissue paper, approximately 0.5 mm on a side

- safety gloves and goggles

A device known as a *fire syringe,* allows a rapid compression of air in a small glass tube that is inside a safety tube of Plexiglas. If pushed very hard, the piston in the glass tube can be forced almost down to the end of the straight-walled section of the tube. If this is done rapidly enough, there is no time for heat energy to be transferred and the compression can be nearly *adiabatic.*

You will either carry out this procedure yourself or have it demonstrated to you.

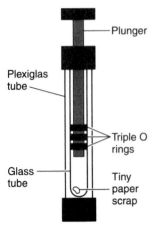

A fire syringe that allows a rapid compression of trapped air to ignite a piece of paper.

1. A small piece of tissue paper should be at the bottom of the syringe, and the piston should be near the top.

2. Hold the syringe vertically on the table. First push the piston in *slowly* as far down as it will go.

3. With the piston at the top of the syringe, measure the initial length of the air column from the bottom of the piston to the bottom of the syringe and record it in Table E2-1.

4. Now, as rapidly as possible, push the piston down as far as it will go while someone holds the ruler nearby and watches carefully to see how far down the bottom of the piston goes. Observe what happens to the paper, and record the approximate final length of the air column when the piston was at its lowest position.

5. Also measure the inside diameter of the fire syringe tube and the initial temperature of the air (room temperature).

Question E2-1: What did you observe happen to the piece of paper? Do you have any explanation for what caused this to happen?

What happens to an ideal gas if it is compressed adiabatically? Work is done on the gas but no heat energy is transferred to the surroundings. Unlike an isothermal compression in which heat energy is transferred out and the internal energy and temperature remain constant, for an adiabatic compression, the internal energy increases and *the temperature increases.* It can be shown mathematically that the relationship between the final and initial temperatures (T_f and T_i) and the fi-

Table E2-1

Initial length of air column (cm)	
Final length of air column (cm)	
Inside diameter of the tube (cm)	
Initial volume V_i (calculated) (cm^3)	
Final volume V_f (calculated) (cm^3)	
Initial temperature T_i (room temperature) (K)	
Final temperature T_f (calculated) (K)	

nal and initial volumes (V_f and V_i) for a diatomic ideal gas (which approximates the behavior of air) undergoing an adiabatic process is given by

$$T_f^{2.5}V_f = T_i^{2.5}V_i$$

(For a monatomic ideal gas, the expression is similar, but the exponent is 1.5 instead of 2.5.) Note that the ideal gas law, $PV = (constant)T$, is still correct, but the fact that the process is adiabatic has put an additional constraint on the system, described by the mathematical equation above.

Now you can do an approximate calculation of the final temperature of the air in the fire syringe when it is compressed so that the piston is as far down as it can go.

Question E2-2: Calculate the approximate final temperature of the air in the fire syringe using the data in your table and the equation for V and T in an adiabatic process and enter it in Table E2-1.

Question E2-3: Compare your calculated final temperature to the "flash point" or burning point of paper—451°F.* Can you now understand why the paper ignited?

Question E2-4: Suppose that a diatomic gas is compressed so that its volume is decreased to 1/10 of its initial value. It started at room temperature, 295 K. What was the final temperature if the process was *adiabatic?*

Question E2-5: What would the final temperature be in Question E2-4 if the process were *isothermal* rather than *adiabatic?*

Question E2-6: Why didn't the tissue paper catch fire when you compressed the air slowly?

Activity 2-2: Work Done by an Expanding Gas

We have examined an isothermal and an adiabatic compression in detail. As you saw in Lab 5, other processes are possible. A process with no pressure change is called *isobaric* and one with no volume change is called *isovolumetric*.

A system we have already met often in our study of thermodynamics is a mass of gas confined in a syringe with a movable piston. The behavior of a gas compressed and expanding in a syringe is a simulation of what goes on in the cylinders in a real engine like the internal combustion engine in your car, or, in fact, even more like a steam engine.

*The paper flash point of 451° is well known to readers of Ray Bradbury's famous science fiction novel, *Fahrenheit 451*, about book burning.

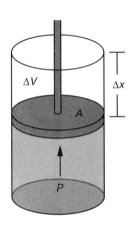

Question 2-7 (review): If a gas expands inside a cylinder with a movable piston so that the volume changes by an amount ΔV while the pressure is kept constant at a value P (isobaric process), what is the mathematical expression to calculate the amount of work done by the gas?

Question 2-8 (review): The graph that follows shows an isobaric expansion from a volume V_a to a volume V_b represented on a P-V diagram. How can you find the work done in the expansion from a to b *from the graph*? (**Hint:** Use your answer to review Question 2-7.)

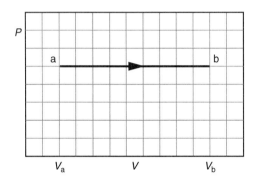

Hopefully, you answered $W = P \, \Delta V$, and that this is just the area under the P vs. V curve. It turns out mathematically that you can always calculate the work done by finding the area under the curve, even if the pressure does not remain constant during the process. Also, the work calculated in this way is *positive* if the gas *expands* in the process. If the gas is compressed during the process, then work is done *on* the gas by the surroundings, and the work done by the system is *negative*. Examples are shown below.

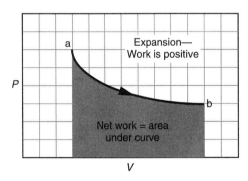
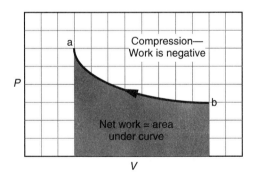

You have already seen in your examination of the first law of thermodynamics in Lab 4 that transferring heat energy to a system can increase its internal energy, but it might result instead in the system doing work on its surroundings. In applying thermodynamics to the operation of heat engines we are interested in the relationships between the heat energy transferred to a system and the work done *by* the system.

According to the first law of thermodynamics, the conservation of energy, including the internal energy, is

$$\Delta E_{\text{int}} = Q - W$$

where Q is the net heat energy transferred to the system (a positive number if heat energy is transferred *into* the system) and W is the work done by the system (a positive number if work is done *by the system*). That is, *transfer of heat energy into the system increases* the internal energy, and *work done by the system decreases* the internal energy. (For a monatomic ideal gas, the internal energy E_{int} is just the sum of all the kinetic energy associated with the random translational motions of molecules.)

One of the key features of our rubber band engine is that it must cool, by transferring heat energy to its surroundings, to be ready to lift the next can. To do so, the surroundings must be cooler than the hot air from the hair dryer. After the rubber band has cooled to its original temperature, and stretched back to its original length, it is in the same *thermodynamic state* that it was in at the start. In other words, all its properties, including its internal energy, are the same. *For one complete cycle of our rubber band engine $\Delta E_{int} = 0$.*

If Q_H is the heat energy transferred to the rubber band from the air heated by the hair dryer and Q_C is the heat energy transferred from the rubber band to the cooler room air, the net heat energy transferred to the rubber band in the cycle is $Q = Q_H - Q_C$ and the first law of thermodynamics becomes

$$\Delta E_{int} = Q - W = (Q_H - Q_C) - W$$

Since $\Delta E_{int} = 0$ for our *complete* cycle, we can simplify this by writing

$$W = Q_H - Q_C$$

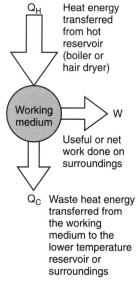

Q_H Heat energy transferred from hot reservoir (boiler or hair dryer)

W Useful or net work done on surroundings

Q_C Waste heat energy transferred from the working medium to the lower temperature reservoir or surroundings

Heat engine schematic.

This basic fact about heat engines is often discussed in terms of an energy flow diagram such as the one shown on the right. This diagram would work equally well for an old fashioned steam engine or our rubber band can lifter.

The figure on the right is a pictorial representation of what we have written in words: Our engine has heat energy Q_H transferred to it, does work W, and transfers some of the original heat energy Q_C to the lower temperature surroundings.

Finding Net Work Done in a Complete *P-V* Cycle

As we have seen, during parts of a cycle when a gas is expanding it is doing *positive* work *on the surroundings*. When it is being compressed, work is being *done on the gas by the surroundings,* so the work done comes out *negative.*

Typically, at the completion of a heat engine cycle, the gas has the same internal energy, temperature, pressure, and volume that it started with. It is then ready to start another cycle. During various phases of the cycle, (1) heat energy transferred to the gas from the hot reservoir (e.g., a boiler) causes the gas to do work on its surroundings as it expands, (2) the surroundings do work on the gas to compress it, and (3) the gas transfers waste heat energy to the surroundings or cold reservoir.

Real heat engines have linkages between a moving piston and the gas or other working medium, which allows the expansion and compression phases of the cycle to run automatically. Thus, some of the work done on the surroundings provides the work needed to compress the gas to return it to its starting point. The useful or net work done in an engine cycle must account for the positive work done during expansion and the negative work done during compression.

Because the work done going from one state to another in one direction is positive and the work done in the other direction is negative, it can be shown mathematically that the work done around a closed loop on a *P-V* diagram, representing a complete cycle of the engine, is the same as the area enclosed by the trace of the process on the diagram. This is illustrated below for two different imaginary cycles.

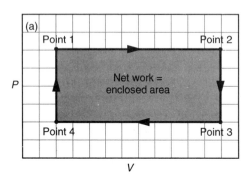
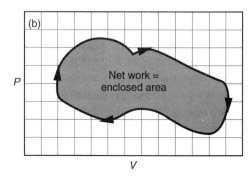

In the next investigation, you will attempt to verify this relationship between useful work and the area on a *P-V* diagram for a real engine.

INVESTIGATION 3: THE INCREDIBLE MASS-LIFTING HEAT ENGINE

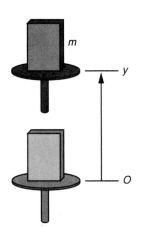

Doing useful mechanical work by lifting a mass *m* through a height *y*.

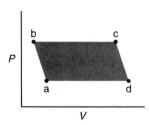

Doing thermodynamic work in a heat engine cycle.

Your working group has been approached by the Newton Apple Company about testing a heat engine that lifts apples that vary in mass from 50 to 100 g from a processing conveyor belt to the packing conveyor belt, which is 5 cm higher. The engine you are to experiment with is a "real" thermal engine that can be taken through a four-stage expansion and compression cycle and that can do useful mechanical work by lifting small masses from one height to another. We would like you to verify experimentally that the useful mechanical work done in lifting a mass *m* through a vertical distance *y* is equal to the net thermodynamic work done during a cycle as determined by finding the enclosed area on a *P-V* diagram. Essentially you are comparing useful mechanical *mgy* work (which we hope you believe in and understand from earlier labs) with the accounting of work in an engine cycle *given by the area enclosed by the cycle*.

Although you can prove mathematically that this relationship holds, the experimental verification will allow you to become familiar with the operation of a real heat engine.

In addition, it will be possible to calculate the heat energy transferred into the heat engine, and compare this to the useful work output. In this way, an approximate value for the efficiency of the engine can be calculated.

To carry out this experiment you will need

- 10-cm³ low-friction glass syringe with ring stand support
- several lengths of Tygon tubing
- 25-mL flask with one-hole rubber stopper
- 2 insulated (e.g., Styrofoam) containers (to use as reservoirs)
- ruler

- 50-g mass

- hot water (about 80–90°C)

- ice water

- computer-based laboratory system with a pressure sensor and a temperature sensor

- data logger software

- *RealTime Physics Heat and Thermodynamics* experiment configuration files

The cylinder of the incredible mass-lifter engine is a low-friction glass syringe. The flat top of the handle of the piston serves as a platform for lifting masses. The flask and pressure sensor can be connected to the syringe with short lengths of flexible Tygon tubing, and the flask can be placed alternately in a cold reservoir and a hot reservoir. A schematic diagram of this mass lifter follows.

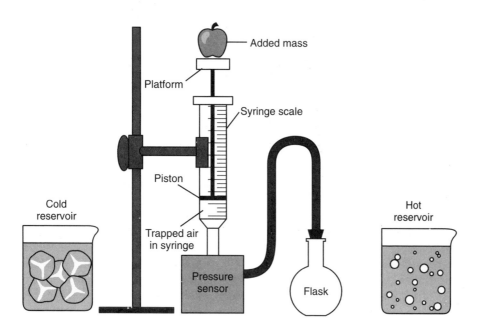

If the temperature of the air trapped inside the cylinder, hose, and flask is increased, then its pressure will increase, causing the platform to rise. Thus, you can increase the volume of the trapped air by moving the flask from the cold to the hot reservoir. Then when the mass has been raised through a distance *y*, it can be removed from the platform. The platform should then rise a bit more as the pressure on the cylinder of gas decreases a bit. Finally, the volume of the gas will decrease when the flask is returned to the cold reservoir. This causes the piston to descend to its original position once again. The various stages of the mass lifter cycle are shown in the diagrams that follow.

The lifting and lowering parts of the cycle should be approximately *isobaric*, since the pressure in the air trapped in the syringe is determined by the weight of the piston (and the mass on top of the handle) pushing down on the gas. The other two parts of the cycle, when the mass is added and removed from the piston handle, should be approximately *adiabatic*, since they occur very quickly.

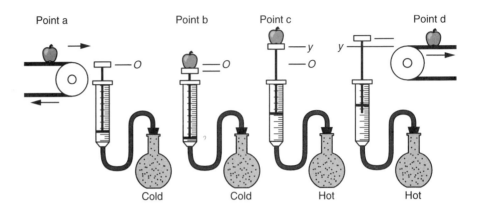

Point a   Point b   Point c   Point d

Cold   Cold   Hot   Hot

Before taking data on the pressure, air volume, and height of lift with the heat engine, you should set it up and run it through a few cycles to get used to its operation. A good way to start is to fill one container with ice water and the other with hot tap water or preheated water at about 80–90°C.

The engine cycle is much easier to describe if you begin with the piston resting above the bottom of the syringe. Thus, we suggest you raise the piston so that the volume of air trapped in the syringe is about 3–4 mL before inserting the rubber stopper firmly in the flask. Also, air does leak out of the syringe slowly. If a large mass is being lifted, the leakage rate increases, so we suggest that you limit the added mass to 50 g.

IMPORTANT: As you take the engine through its cycle, observe whether the piston is moving freely in the syringe. If it is sticking, it should be removed and dipped into *distilled* water to free it up. If it continues to get stuck, ask your instructor for help.

After observing a few engine cycles, you should be able to describe each of the points **a, b, c,** and **d** of a cycle, carefully indicating which of the transitions between points are approximately adiabatic and which are isobaric.

You should reflect on your observations by answering the questions in the next activity. You can observe changes in the volume of the gas directly and you can predict how the pressure exerted on the gas by its surroundings ought to change from point to point by using the definition of pressure as force per unit area.

Activity 3-1: Description of the Engine Cycle

Prediction 3-1: With the system closed to the outside air and the flask in the cold reservoir, what should happen to the height of the platform during transition **a→b,** as you add the mass to the platform? Explain the basis of your prediction.

1. Make sure the rubber stopper is firmly in place in the flask. Add the mass to the platform.

Question 3-1: Describe what happened. Is this what you predicted? Why might this process be approximately adiabatic?

Prediction 3-2: What do you expect to happen during transition **b→c,** when you place the flask in the hot reservoir?

2. Place the flask in the hot reservoir. (This is the engine power stroke!)

Question 3-2: Describe what happens. Is this what you predicted? Why should this process be isobaric?

Prediction 3-3: If you continue to hold the flask in the hot reservoir, what will happen when the added mass is now lifted and removed from the platform during transition **c→d** (and moved onto an upper conveyor belt)? Explain the reasons for your prediction.

3. Remove the added mass.

Question 3-3: Describe what actually happens. Is this what you predicted? Why might this process be approximately adiabatic?

Prediction 3-4: What do you predict will happen during transition **d→a,** when you now place the flask back in the cold reservoir? Explain the reasons for your prediction.

4. Now it's time to complete the cycle by cooling the system down to its original temperature for a minute or two before placing a new mass to be lifted on it. Place the flask in the cold reservoir.

Question 3-4: Describe what actually happens to the volume of the trapped air. Why should this process be isobaric?

Question 3-5: How does the volume of the gas actually compare to the original volume of the trapped air at point **a** at the beginning of the cycle? Is it the same or has some of the air leaked out?

Question 3-6: Theoretically, the pressure of the gas should be the same once you cool the system back to its original temperature. Why?

Table 3-2a

State of system	Volume of air in syringe (cm^3)	Volume of flask (cm^3)	Volume of tubing (cm^3)	Volume of sensor (cm^3)	Total volume of air (cm^3)
a		25			
b		25			
c		25			
d		25			
a'		25			

To calculate the thermodynamic work done during a cycle of this engine you will need to be able to plot a *P-V* diagram for the engine based on determinations of the volumes and pressures of the trapped air in the cylinder, Tygon tubing, and flask at the points **a, b, c,** and **d** in the cycle. You can do this by hand, or you can have your computer-based system do it for you.

Activity 3-2: Work Done by the Heat Engine

1. Estimate the total volume of the tubing between the flask and syringe and the pressure sensor using the inside diameter and length. Show your calculation.

Inside diameter:_____cm   Total length:_____cm

Estimated volume of tubing:_____

Enter this value in the fourth column of Table 3-2a.

2. Enter the volume of the pressure sensor given to you by your instructor in the fifth column of Table 3-2a.

3. Connect the pressure sensor and temperature sensor to the interface and start up the software.

4. Open the experiment file called **Pressure and Temperature (L6A3-2).** This will display the axes that follow for pressure vs. volume. This will also set up the software in **prompted event mode** so that you can continuously measure pressure and decide when you want to **keep** a value. Then you can **enter** the measured volume.

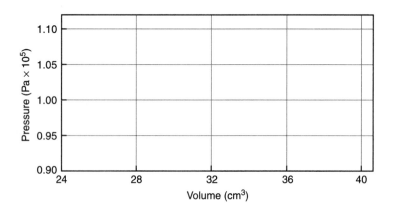

5. **Load the calibration files** for the temperature and pressure sensors.

6. Record the value of the mass to be lifted in Table 3-2b.

Now you should be able to take your engine through another cycle and make the measurements of volume and pressure of the air needed to determine the *P-V* diagram for your heat engine. You should take your data *rapidly* to avoid air leakage around the piston.

7. Begin with the flask and temperature sensor in the ice water, and without the mass on the handle of the syringe (state **a**). Stir the ice water. **Begin collecting data.** When the temperature and pressure seem to be fairly stable, **keep those data values.**

8. Read the volume of air in the syringe, enter it in Table 3-2a, calculate the total volume of air, and **enter this value** into the computer.

9. *Quickly* place the mass on top of the handle of the syringe (state **b**).

10. When the temperature and pressure seem to be fairly stable, **keep those data values.** Again, record the volume of air in the syringe in the table, calculate the total volume of air, and **enter this value** into the computer.

11. Quickly move the flask and temperature sensor to the hot water reservoir (state **c**). When the temperature and pressure seem to be fairly stable, **keep those data values.** Again, record the volume of air in the syringe in the table, calculate the total volume of air, and **enter this value** into the computer.

12. Quickly remove the mass (state **d**). When the temperature and pressure seem to be fairly stable, **keep those data values.** Again, record the volume of air in the syringe in the table, calculate the total volume of air, and **enter this value** into the computer.

13. Finally, move the flask and temperature sensor back to the ice-water reservoir (state **a'**). Stir the ice water. When the temperature and pressure seem to be fairly stable, **keep those data values.** Again, record the volume of air in the syringe in the table, calculate the total volume of air, and **enter this value** into the computer.

14. Measure the height that the mass was raised. This can easily be done after all measurements by going back, looking at your volume data and measuring the difference in positions of the piston from state **b** to state **c**. Record in Table 3-2b.

15. Read the temperatures of the two water reservoirs (states **a** and **c**) from the data table and record them in Table 3-2b.

16. **Print the graph and the data table** and affix them over the previous axes.

Table 3-2b

Mass to be lifted (g)	
Height mass was raised y (m)	
Hot reservoir temperature (K)	
Cold reservoir temperature (K)	

Question 3-7: You expected that the transitions from **b**→**c** and from **d**→**a** were isobaric. According to your data, were they? Explain.

Activity 3-3: Calculating the Work Done by the Heat Engine

There are three ways to find the area of the cycle that gives the work done by the heat engine:

Method I: Use the **integration routine** in the software to find the area.

Method II: Since the pressure doesn't change (much) from point **b** to point **c** you can take the average pressure of those two points as a constant pressure. The same holds for the transition from **d** to **a**. This gives you a figure that is approximately a parallelogram with two sets of parallel sides. You can look up and properly apply the appropriate equation to determine the area and net thermodynamic work performed.

Method III: Display your graph with a grid and count the boxes in the area enclosed by the lines connecting points **a**, **b**, **c**, and **d.** Then multiply by the number of joules each box represents. You will need to make careful estimates of fractions of a box when a cycle cuts through a box.

Find the work in joules (J) by one of these methods. Show all work below. Be sure that your units are correct. (Pressure will need to be in Pa and volume in m^3.)

If you use the computer method, you will need to be careful if the cycle did not close on itself. You should measure the work done in each part of the cycle (**a**→**b**, **b**→**c** , **etc.**) and combine these together.

If you use Method III, use the axes below. Fill in the scales.

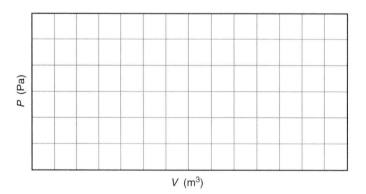

1. Work done by heat engine. Show all calculations below.

2. Use the equation $W = mgy$ to calculate the useful mechanical work done in lifting the mass from one level to the other in joules (J).

Question 3-8: How does the thermodynamic work compare to the useful mechanical work? Please use the correct number of significant figures in your comparison (as you have been doing all along, right?)

If you have time, carry out Extension 3-3 to calculate the efficiency of your heat engine.

Extension 3-4: Efficiency of the Mass-Lifting Heat Engine

The efficiency of a heat engine is defined in the following way:

$$e = 100\% \left(\frac{W_{out}}{Q_{in}} \right) = 100\% \left(\frac{W}{Q_H} \right)$$

You have just found W. The heat energy input from the hot reservoir takes place in the process **b→c**. (Remember that **c→d** is an *adiabatic process with no heat energy transfer.*)

The heat energy transferred into a gas during an isobaric process in which the temperature changes by ΔT is given by

$$Q = nC_P \, \Delta T$$

where n is the number of moles of gas and C_P is the molar heat capacity at constant pressure, which is 29.0 J/mol-K for air.

The most efficient possible heat engine operating with a hot reservoir at T_H and a cold reservoir at T_C is called a Carnot engine, after Sadi Carnot, the French engineer who studied engine efficiencies in the early nineteenth century. According to his theoretical calculations, the maximum possible, or Carnot, efficiency is given by

$$e_c = 100\% \left(1 - \frac{T_C}{T_H} \right)$$

where both T_C and T_H are in K.

In the following activity you will determine the efficiency of your engine and of a Carnot engine operating between the same two reservoirs.

1. Calculate the number of moles of gas in your system. (**Hints:** Use the ideal gas law $PV = nRT$ and your data for state **a**, with P in Pa, V in m^3, and T in K, $R = 8.314$ J/mol-K.)

2. Calculate the heat energy transferred into the gas during the process **b→c**. (**Hint:** Use the equation above for Q, and the temperatures of the two reservoirs.)

3. Calculate the efficiency of the mass-lifting heat engine.

4. Calculate the efficiency of a Carnot engine operating with the same hot and cold reservoirs.

Question E3-10:   Is the mass-lifting heat engine very efficient? What percentage of the input heat energy is converted to useful work? What percentage is lost as waste heat energy?

Question E3-11:   How does the efficiency of the mass-lifting heat engine compare to the maximum possible efficiency (the Carnot efficiency)? Are you surprised by the answer?

Comment: Note that the incredible mass-lifting engine is actually not so simple. Understanding the stages of the engine cycle on a *P-V* diagram is reasonably straightforward. However, it is difficult to use equations for adiabatic expansion and compression and the ideal gas law to determine the temperature (and hence the internal energy) of the air throughout the cycle. There are several reasons for this. First, air is not an ideal gas. Second, the mass-lifting engine is not well insulated and so the air that is warmed in the hot reservoir transfers heat energy through the cylinder walls. Thus, the air in the flask and the air in the cylinder are probably not at the same temperature. Third, air does leak out around the piston, especially when larger masses are added to the platform. This means that the number of moles of air decreases over time. (You can observe this by noting that in the transition from point **d** to point **a** the piston can actually end up in a lower position than it had at the beginning of the previous cycle.) However, the incredible mass-lifting engine does help us understand typical stages of operation of a real heat engine.

HOMEWORK FOR LAB 6:
HEAT ENGINES

1. If 5.0 moles of air is at 0.0°C and 1.0 atm pressure (otherwise known as standard temperature and pressure, or STP for short), what is the volume occupied by the gas in m^3? (Treat air as an ideal gas.)

2. Suppose that the gas in Question 1 is compressed *isothermally* until its volume is one-fifth as large. What is the final pressure? What is the final temperature?

3. The air in Question 1 is instead compressed *adiabatically* until its volume is one-fifth as large. What is the final temperature? What is the final pressure?

4. Explain macroscopically in terms of work and energy considerations why the temperatures in Questions 2 and 3 are different. What happens to the internal energy during the process in Question 2? What happens to the internal energy during the process in Question 3?

5. On a microscopic level, how does the motion of molecules change when the temperature of a gas increases? Explain microscopically, in terms of collisions of gas molecules with the rapidly moving piston, why the temperature of a gas rises in an adiabatic compression.

6. The air in Question 1 is compressed *isobarically* until its volume is one-fifth as large. How much work was done? Is the work done positive or negative?

7. Sketch the process in Question 6 on the axes below and explain how to calculate the work done from the graph.

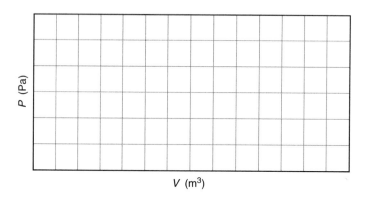

8. Calculate the work done in the process from state 1 to state 2 shown on the P-V diagram below. (**Note:** 1 MPa = 10^6 Pa.) Is the work positive or negative? Explain. What would the work be if the process went from 2 to 1?

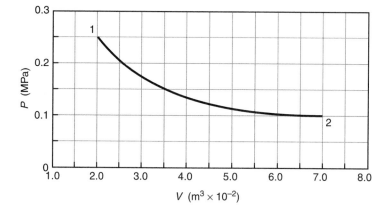

9. Suppose 1.5 moles of air are carried through the cyclic process shown below. Calculate the work done during the cycle. (Note: 1 MPa = 10^6 Pa.) Is the work positive or negative? Explain. What would the work be if the cycle were carried out in the counterclockwise direction?

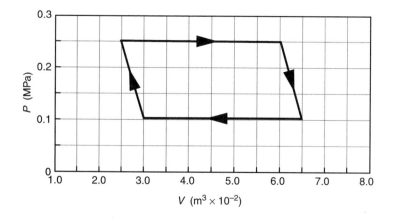

10. Calculate the net heat energy transfer into the system for the cycle in Question 9.

11. Assuming that the process in Question 9 consists of two isobaric and two adiabatic processes, find the net heat energy input to the system.

12. Find the efficiency of the cycle in Question 9.

13. Find the efficiency of a Carnot engine operating between the highest and lowest temperatures in the cycle in Question 9.

Appendix A:
Realtime Physics
Heat and Thermodynamics
Experiment Configuration Files

Listed below are the settings in the *Experiment Configuration Files* used in these labs. These files are available from Vernier Software for the compatible software which they sell (*MacTemp, Temperature*—for MS-DOS, *Data Logger* and *Logger Pro*—for Windows,) and from PASCO Scientific for *Science Workshop*. They are listed here so that the user can set up files for any compatible hardware and software package.

Experiment File	Description	Data Collection	Data Handling	Analysis	Display
Digital Readouts (L1A1-1)	Displays digital readings of temp. sensors 1 and 2	10 points/s Temp. sensors 1 and 2 Digital display of inputs	NA	NA	Digital display Temp.: 0–100°C
Reaching Equilibrium (L1A1-2)	Displays and graphs temp. sensors 1 and 2 vs. time	10 points/s Temp. sensors 1 and 2 Digital display of inputs	NA	NA	One set of graph axes with lines Temp.: 0–30°C Time: 0–240 s
Cooling Water (L1A2-1)	Displays temp. sensors 1 and 2, and graphs temp. sensor 1 vs. time	10 points/s Temp. sensors 1 and 2 Digital display of inputs	NA	NA	One set of graph axes with lines Temp.: 0–100°C Time: 0–10 min
Temp. Change and Heat Transfer (L1A2-2)	Displays and graphs temp. sensors 1 and 2 vs. time	10 points/s Temp. sensors 1 and 2 Digital display of inputs	NA	NA	One set of graph axes with lines Temp.: 0–100°C Time: 0–5 min
Keeping Hot (L1A3-1)	Displays and graphs temp. sensor 1 only vs. time Enables heat pulser to transfer heat pulses	10 points/s Temp. sensor 1 only Digital display of input Heat pulser enabled, 2 s pulse length	NA	NA	One set of graph axes with line Temp.: 70–90°C Time: 0–120 s

Experiment File	Description	Data Collection	Data Handling	Analysis	Display
Heating Water (L1E3-2)	Displays and graphs temp. sensor 1 only vs. time Enables heat pulser to transfer heat pulses	10 points/s Temp. sensor 1 only Digital display of input Heat pulser enabled, 5 s pulse length	Use to display data from first part persistently on the screen for comparison with data from second part	May use analysis feature to read data from graphs after collected	One set of graph axes with line Temp.: 10–40°C Time: 0–150 s
Mech. and Elect. Work (L2A1-1)	Displays and graphs temp. sensor 1 only vs. time	10 points/s Temp. sensor 1 only Digital display of input	Use to display data from first part persistently on the screen for comparison with data from second part	May use analysis feature to read data from graphs after collected	One set of graph axes with line Temp.: 20–40°C Time: 0–180 s
Heating Water (L2A2-1)	Displays and graphs temp. sensor 1 only vs. time Enables heat pulser to transfer heat pulses	10 points/s Temp. sensor 1 only Digital display of input Heat pulser enabled, 5 s pulse length	Use to display data from first part persistently on the screen for comparison with data from second part	May use analysis feature to read data from graphs after collected	One set of graph axes with line Temp.: 20–40°C Time: 0–180 s
Keeping Hot (L3A1-1)	Displays and graphs temp. sensor 1 only vs. time Enables heat pulser to transfer heat pulses	10 points/s Temp. sensor 1 only Digital display of input Heat pulser enabled, 2 s pulse length	NA	NA	One set of graph axes with line Temp.: 70–90°C Time: 0–5 min
Math. Relation (L3E1-2)	Displays data table and graph for number of pulses and temperature difference data which are entered manually	NA	New manual columns of data for # Heat Pulses and Temp. Diff.	Use graphical fit routine to find experimental relationship between number of heat pulses and temp. diff.	One set of graph axes with line # Heat Pulses: 0–20 Temp. Diff.: 0–80°C
Cooling Down (L3A2-1)	Displays and graphs temp. sensors 1 and 2 vs. time	10 points/s Temp. sensors 1 and 2 Digital display of inputs	NA	May use analysis feature to read data from graphs after collected	One set of graph axes with lines Temp.: 40–90°C Time: 0–5 min

Experiment File	Description	Data Collection	Data Handling	Analysis	Display
Radiation (L3A3-2)	Displays and graphs temp. sensor 1 only vs. time	10 points/s Temp. sensor 1 only Digital display of input	Use to display data from first part persistently on the screen for comparison with data from second part	May use analysis feature to read data from graphs after collected	One set of graph axes with lines Temp.: 20–35°C Time: 0–5 min
Ice to Water (L4A1-1)	Displays and graphs temp. sensor 1 only vs. time Enables heat pulser to transfer heat pulses	10 points/s Temp. sensor 1 only Digital display of input Heat pulser enabled, 10 s pulse length	NA	May use analysis feature to read data from graphs after collected	One set of graph axes with line Temp.: −10–30°C Time: 0–5 min
Water to Steam (L4A2-1)	Displays and graphs temp. sensor 1 only vs. time Enables heat pulser to transfer heat pulses	5 points/s Temp. sensor 1 only Digital display of input Heat pulser enabled, 10 s pulse length	NA	May use analysis feature to read data from graphs after collected	One set of graph axes with line Temp.: 10–110°C Time: 0–10 min
Measuring Pressure (L5A1-2)	Displays pressure sensor 1 only	10 points/s Pressure sensor 1 only Digital display of input	NA	NA	No axes. Just digital display
Pressure vs. Volume (L5A2-1)	Displays pressure and volume Pressure is continuously measured while volume is entered manually after pressure value is kept	10 points/s Pressure sensor 1 only Digital display of inputs Prompted event mode—keep pressure values only when desired	New column that displays pressure in 10^5 Pa [P(atm) × 1.01325] New prompted manual column to enter volume in cm^3	Use graphical fit routine to find experimental relationship between pressure and volume	One set of graph axes with line Press.: 0–8 × 10^5 Pa Volume: 0–24 cm^3

Experiment File	Description	Data Collection	Data Handling	Analysis	Display
Measuring P and T (L5A2-2)	Displays pressure and temperature Both are continuously measured and can be kept as desired	10 points/s Pressure sensor 1 and temp. sensor 2 Digital display of inputs Event mode—keep pressure and temp. values only when desired	New column that displays pressure in 10^5 Pa [P(atm) $\times$ 1.01325] New column that displays temp. in K [T(°C) + 273.15]	Use graphical fit routine to find experimental relationship between pressure and temp.	One set of graph axes with line Press.: 0.6–1.2 $\times$ 10^5 Pa Temp.: 270–370 K
Measuring V and T (L5A2-3)	Displays volume and temperature Temperature and pressure are continuously measured while volume is entered manually after these values are kept	10 points/s Pressure sensor 1 and temp. sensor 2 Digital display of inputs Prompted event mode—keep pressure and temp. values only when desired	New column that displays pressure in 10^5 Pa [P(atm) $\times$ 1.01325] New column that displays temp. in K [T(°C) + 273.15] New prompted manual column to enter volume in cm^3	Use graphical fit routine to find experimental relationship between volume and temp.	One set of graph axes with line Volume: 25–55 cm^3 Temp.: 270–320 K
Single Particle Movie (L5E3-1)	Digital movie that shows a single particle in a box. Need movie viewer software like Apple Movie Player to view	NA	NA	NA	NA
Syringe Movie (L5E3-2)	Digital movie that shows particles in a syringe with a moving plunger. Need movie view software like Apple Movie Player to view	NA	NA	NA	NA
Pressure and Temperature (L6A3-2)	Displays pressure, temperature and volume Pressure and temperature are continuously measured while volume is entered manually after pressure value is kept	10 points/s Pressure sensor 1 and temperature sensor 2 Digital display of inputs Prompted event mode—keep pressure values only when desired	New column that displays pressure in 10^5 Pa [P(atm) $\times$ 1.01325] New column that displays temp. in K [T(°C) + 273.15] New prompted manual column to enter volume in cm^3	Use graphical fit routine to find area enclosed in pressure vs. volume cycle	One set of graph axes with line Press.: 0.9–1.1 $\times$ 10^5 Pa Volume: 24–40 cm^3